API Analytics for Product Managers

Understand key API metrics that can help you grow your business

Deepa Goyal

BIRMINGHAM—MUMBAI

API Analytics for Product Managers

Group Product Manager: Alok Dhuri
Contributor: Kevin Swiber
Publishing Product Manager: Akshay Dani
Senior Editor: Nithya Sadanandan
Technical Editor: Jubit Pincy
Copy Editor: Safis Editing
Project Manager: Prajakta Naik
Project Coordinator: Manisha Singh
Proofreader: Safis Editing
Indexer: Pratik Shirodkar
Production Designer: Prashant Ghare
Developer Relations Marketing Executive: Deepak Kumar and Rayyan Khan
Business Development Executive: Puneet Kaur

First published: February 2023

Production reference: 1010223

Published by Packt Publishing Ltd.
Livery Place
35 Livery Street
Birmingham
B3 2PB, UK.

ISBN 978-1-80324-765-6

www.packtpub.com

To my parents, Kailash and Renu Goyal, for exemplifying courage, boldness, and humility. To my husband, Sushant Thakur, for his love, encouragement, and support.

– Deepa Goyal

Foreword

I first met Deepa in January of 2021 when she was a guest on my Breaking Changes podcast. I was captivated by the way she described her journey to becoming an API product manager, and how hard it was for her to obtain the knowledge she needed, but also how methodical she was about how she did her research and went about gathering the data and knowledge she needed. Her experience with API product management at relevant API providers such as PayPal and Twilio made her a sure bet for me when it came to launching season two of my podcast.

Fast-forward six months. I had done over 30 other interviews with enterprise organizations about their priorities when it comes to API operations. I spoke with companies such as 7-Eleven, NBA, Boy Scouts of America, Plaid, and many, many more. The top business priority I heard across these conversations was consistently centered around managing your APIs as a product. These podcast episodes also echoed what I was hearing in the conversations I was having with Postman customers—I was hearing from mainstream companies such as Nationwide, HSBC, Zoom, and others about their investment in API product management as a discipline. I knew at that moment that I needed to hire Deepa and put her in charge of helping me understand this fast-growing dimension of the API economy.

Deepa comes to the table with a deep understanding of what API-as-a-product means in an age where most people just possess a superficial marketing awareness that has been shared by leading API management providers over the last decade. Deepa goes beyond just talking about what it means to treat your APIs as a product and can speak to the details and nuance of what API product managers need to be doing on a daily basis. Deepa has been on the ground floor at leading API providers, doing the work, but then also working with me to understand this API product management bridge that has emerged between enterprise business, IT, and platform groups as part of Postman Open Technologies.

Deepa possesses the blueprints for an essential bridge that has emerged over the last decade between historically divided business and IT groups within enterprises. She has what you need when it comes to building, growing, iterating, and supporting your API operations through the adoption of a healthy product mindset. In this book, Deepa provides you with what you need to bridge between your business and IT groups, but also between producers and consumers, whether that is internally, with your partners, or publicly via third-party consumers. For me, this is the trifecta I've been hearing about from leading enterprise organizations who are further along in their API journey—the ability to strike a balance between business and IT, but also across internal and external consumers of your digital products.

I was sold on Deepa's empathy for consumers, and robust understanding of the API landscape and API products from early on. But it was her pragmatic and detailed approach to not just defining what success looks like in this product realm but actually measuring it in meaningful ways. Deepa's approach provides the nutrients and purpose that API product managers will need when it comes to measuring both the technical and business dimensions of the APIs they own. This approach provides you with all the metrics and surrounding scaffolding you will need to scale, evolve, and adapt your approach to delivering APIs, standardizing quality and consistency across APIs and teams. She equips us all with what we need to be successful with the API products we own while helping us take care of the bigger-picture strategy that will matter to leadership.

Buckle up—Deepa is going to give you a professionally guided tour across the API landscape. She is going to make sure you understand both the producer and consumer sides of things in such a balanced way you won't be able to forget about your developers you'll understand them that well. Then, Deepa will help you bring things home by providing you with the metrics you need to track your progress and quantify success while you develop, iterate, and adapt your strategy for shaping your organization's digital products. She is going to provide you with what you need to hit the ground running, but then also ensure you are properly investing in the big-picture strategy and storytelling you will need to lead your team in the right direction.

I get APIs—I have been immersed in them since 2010. However, I come from the IT side of things, so I don't always see the business dimension properly. I also tend to lean toward being an API producer first and an API consumer second. After 6 months of working with Deepa, I find myself regularly reminded of my bias as part of our conversations and her persistent questioning. She is always stopping me and reminding me to talk to customers. She is always shining a light on how entrenched I am when it comes to my API producer mindset. She has opened up the importance of the role that API product managers are playing when it comes to bridging the business and IT divide, increasing the velocity of feedback loops with consumers, and ensuring the APIs we deliver are meeting the needs of the business, but also the customers we serve.

Thanks for writing this book, Deepa—I am confident that I will be working with the material you have provided for the next 5 to 10 years! Or at least until you get the 2nd edition ready. No pressure!

Kin Lane

Chief Evangelist, Postman Inc.

Author of The API-First Transformation

www.APIEvangelist.com

Contributors

About the author

Deepa Goyal is a Silicon Valley veteran with diverse experience in Fortune 500 tech and start-ups. She is currently the lead product management strategist at Postman, where she helps millions of developers build **application programming interfaces** (**APIs**). Previously, she worked at PayPal and Twilio, where she grew the companies' API offerings, which are used by the software developer community for thousands of third-party websites and mobile apps. Deepa is skilled in both product management and data science. She uses the insights she gets from data science in her product management work.

First and foremost, I want to thank my husband, Sushant Thakur, for inspiring me to take on this project. I want to thank my mentor, Kin Lane, for guiding me in shaping this book and providing resources to enrich my understanding of the topics I present. I also want to thank my team at Postman, particularly Kevin Swiber, Pascal Heus, and Arnaud Lauret, for being great sounding boards for ideas and sharing their expertise.

Kevin Swiber was a significant contributor to the book. Here's a little about him:

Kevin Swiber is an API Lifecycle Integration Specialist at Postman and is a leading voice in open technologies. Kevin is a software engineering, architecture, and developer tools advocate, having focused on distributed systems and APIs for over a decade. Their career has taken them from the enterprise to the API management space. Kevin has worked at a number of API product companies, including Apigee (acquired by Google Cloud Platform) and now Postman, where they collaborate with both industry and enterprise companies on building a platform strategy across the entire API lifecycle. Kevin also serves as the marketing chair for the OpenAPI Initiative.

About the reviewers

Sri Kandikonda is a digital product leader with demonstrated results in leading design and strategy for complex B2B technology products (SaaS, machine learning, APIs, platform products) across fintech and e-commerce.

Sri has successfully managed the complexities of global B2B SaaS products (1B+ transactional volume/day, $15B+ in annual revenue) for Fortune 100 companies as well as high-growth start-ups. Sri operates in the intersection of technology, business, and customer experience to create world-class, scalable products that customers love.

Sri has Led large-scale product rollouts and global marketing strategies across the Americas, the EU, and the AMEA market.

Suvrat Gupta is a product leader with many years of experience leading API products across some of the biggest technology platforms, such as Amazon, BNY Mellon, and Dun & Bradstreet. Suvrat has worked extensively with the challenge of establishing API governance at scale, improving the API developer experience, and establishing API analytics to unlock growth across internal as well as external APIs with millions of users across the world. He has led API-driven digital transformation programs in multiple industries and is passionate about APIs as first-class products.

Table of Contents

9

Components of API Experience 133

Part 3: Deep Dive into Key Metrics for API Products

10

Infrastructure Metrics 163

11

API Product Metrics 181

12

Business Metrics 209

Part 4: Setting a Cohesive Analytics Strategy

13

Drawing the Big Picture with Data 235

14

Keeping Metrics Honest 259

15

16

Preface

Application programming interfaces (APIs) have become a ubiquitous part of web technologies because they provide a standard way for different systems and applications to communicate and share data and functionality. This allows for greater flexibility, scalability, innovation, and interoperability in the web ecosystem.

As more and more organizations build APIs for use both inside and outside the company, different areas of design, documentation, governance, and life cycle have become more important. However, until now, APIs have only been viewed as technical components. I want to change the way APIs are thought of by seeing them as fully qualified products.

Because "API as a product" refers to a "product," we consider an application program to be a standalone product rather than a technical component of a larger system. This means that an API is developed, marketed, and supported in a similar way as other products, with a focus on meeting the needs of specific user groups and delivering value to them.

The "API as a product" approach lets companies sell their data and services, while also giving developers a useful tool for making new apps and services. APIs as products also have revenue-generation potential, as they can be used as products themselves, either by licensing them to a third party or charging for their usage by end users.

I have organized this methodology into four key areas:

- How to manage APIs as products
- How to build customer empathy for API products
- How to design metrics for measuring APIs from the infrastructure, product, and business perspectives
- How to identify the right **key performance indicators** (**KPIs**) for API products and build a product strategy

This book will guide you through the process of managing APIs as products, building customer empathy, designing metrics for measuring success, and identifying KPIs to inform your product strategy. Whether you are new to the field or a seasoned professional, this book will provide valuable insights and best practices for optimizing the performance and profitability of your API products. Let's dive in and discover how to unlock the full potential of your APIs.

Who this book is for

This book is written for product managers, business leaders, and developers who are looking to get the most out of their APIs. It gives an in-depth look at the most important ideas and best practices related to API analytics and product management. This makes it an essential resource for anyone who builds, deploys, or manages API products.

Additionally, this book serves as a valuable resource for security teams, sales teams, operations personnel, and user experience researchers who are involved with APIs. These teams will benefit from the book's detailed guidance on how to design metrics for measuring API performance, as well as the strategies for understanding user behavior and feedback that can inform the design of more scalable, secure, and user-friendly APIs.

What this book covers

Part 1, The API Landscape

The objective of this part is to introduce APIs as products and shed light on how large the market is for API products. You will learn about product management concepts and how they apply to APIs. This part will also explain the life cycle and maturity of an API.

Chapter 1, API as a Product

APIs go beyond web products or mobile apps with the UI. In this chapter, you will be introduced to the idea of an API as a product and how a vast universe of products is built using APIs. This chapter will also look at some of the most well-known API companies and how they've made successful API products.

Chapter 2, API Product Management

API product management has evolved into a specialization with some fundamental pieces that a product manager must understand to effectively make product decisions. This chapter will go over various types of products from a product management perspective and how they require different skill sets.

Chapter 3, API Life Cycle and Maturity

This chapter will help you understand why the API product life cycle, methodology for establishing API governance, and use of the API maturity model are important for organizations, as they help them to ensure that their APIs are developed and managed in a consistent, efficient, and effective manner, aligned with the organization's goals, policies, and standards, and that they can evolve over time to meet changing business needs. This chapter also presents case studies of some of the leading API products and how they present their API maturity to their customers.

Chapter 4, Building and Managing API Products

This chapter will talk about the unique design challenge of defining an API product MVP. As the API product matures, the challenges can get more complicated, and in addition to growth, retention and churn might also become very crucial in product strategy. At each step of API maturity, the stakeholders' and customers' needs and expectations change. This chapter explains what we mean by "API maturity" and how it relates to the API life cycle.

Chapter 5, Growth for API Products

Growth for APIs refers to the process of increasing the usage and adoption of an API by different user groups, such as developers, businesses, and consumers. Growth can be achieved by identifying, helping identify, and helping the target audience; developing a marketing, pricing, and sales strategy that effectively communicates the value and benefits of the API to the target audience; and helping to generate interest and awareness. We can utilize product-led growth and community-led growth for API growth.

Chapter 6, Support Models for API Products

The customer support strategy for API products is different from that for other products. This chapter dives into the standard methodologies for creating a robust support model for APIs that scales with the product and delivers value for customers.

Part 2, Understanding the Developer

This part is focused on the primary customer of APIs: the developer. It is evidently important to understand the developer journey in order to establish a growth funnel for your API product. You will also learn about signals for activation, engagement, retention, and scale.

Chapter 7, Walking in the Customer's Shoes

This chapter describes what product funnels are and how they are established for various types of products. You will be introduced to concepts such as activation, retention, engagement, and churn.

Chapter 8, Customer Expectations and Goals

This chapter helps you understand the goals of both the business and the customer to be able to establish roadmaps that build a long-term API strategy for the organization. This chapter will introduce you to tools such as CSAT, NPS, and other user research mechanisms to develop an understanding of customers. You will learn how to understand your customers so you can get them to use your product, and set up a product strategy that gets customers started on a long-term relationship with your product.

Chapter 9, Components of API Experience

In this chapter, you will learn about a few key ingredients for creating a great API experience. It is important to understand how some of these experiences have been designed across the industry to be able to shape any API product. We look at things such as API references, status pages, SDKs, CLIs, and so on that are part of the API experience.

Part 3, A Deep Dive into Key Metrics for API Products

This part will introduce you to the reasoning behind API metrics. You will do a deep dive into all dimensions of the user journey and learn about a vast set of metrics that you can track across the infrastructure, product, and business dimensions of your APIs.

Chapter 10, Infrastructure Metrics

Infrastructure metrics are crucial for APIs that serve a large or a small customer base. It is important that APIs be reliable. In this chapter, you will learn how to measure infrastructure metrics and various tools that provide an easy setup to get them.

Chapter 11, API Product Metrics

In this chapter, you'll find out about the different metrics you can use to learn more about your customers. The metrics you learn in this chapter can be used across all the stakeholders in your product to align on common goals and priorities.

Chapter 12, Business Metrics

In this chapter, you'll learn about the business metrics you need to set up and keep track of regularly in order to measure the business impact of your infrastructure and product development projects.

Part 4, Setting a Cohesive Analytics Strategy

It is not sufficient to merely have metrics set up. It is also important to understand how to evaluate the quality of the metrics and how to make sure they are extensive and robust. This part describes the possible ways in which metrics can be analyzed and evaluated. You will learn how to remove blind spots and avoid vanity metrics that may not be true representations of product health.

Chapter 13, Drawing the Big Picture with Data

This chapter dives into the evaluation of metrics once a measurement is done. The first step is to establish a baseline and find ways of benchmarking it. Metrics should not be standalone; they need to be evaluated in the context of other metrics. This chapter also establishes the concept of correlation in metrics and dives into how to set clusters of metrics so that there is a set of metrics that are seen in relation to each other and not all metrics at once.

Chapter 14, Keeping Metrics Honest

This chapter talks about combining qualitative and quantitative data to form hypotheses and drive insights that may not be easily available without combining these two. This chapter also explains what leading and lagging metrics are and how to find them in a set of related metrics.

Chapter 15, Counter Metrics to Avoid Blind Spots

In this chapter, you will learn about counter metrics to remove bias from the metrics-setting process so that blind spots might be addressed. This chapter also introduces the concept of gamaebility with examples and explains the consequences of gameable and vanity metrics.

Chapter 16, Decision-Making with Data

In this chapter, you will learn about how effective product leadership requires setting short-term and long-term goals and strategically communicating those goals to stakeholders through storytelling. This approach helps to establish a clear direction for the product and the team, aligning everyone around a common vision and enabling the team to work together to achieve success.

To get the most out of this book

It is recommended that you have a good understanding of web development, software development, product management, and data analysis to get the most out of this book. The prerequisites include the following:

- A basic understanding of what APIs are and how they work
- Familiarity with programming concepts and experience with web development would be beneficial, as would an understanding of the software development life cycle and product management concepts
- It would be helpful for you to have some prior experience with data analysis and an understanding of KPIs, as the book covers how to design metrics for measuring APIs from the infrastructure, product, and business perspectives, and how to identify the right KPIs for API products and build a product strategy
- Additionally, it would be beneficial for you to have some understanding of user experience research and user-centered design, as the book covers how to build customer empathy for API products and how to use customer research to inform product development

There is no special software installation required for this book.

Download the color images

We also provide a PDF file that has color images of the screenshots and diagrams used in this book. You can download it here: `https://packt.link/sQ5oJ`.

Conventions used

There are a number of text conventions used throughout this book.

Code in text: Indicates code words in text, database table names, folder names, filenames, file extensions, pathnames, dummy URLs, user input, and Twitter handles. Here is an example: "Each path can have a defined `GET/PUT/POST/DELETE HTTP` action and may have predetermined conditions."

Bold: Indicates a new term, an important word, or words that you see onscreen. For instance, words in menus or dialog boxes appear in bold. Here is an example: "Select **System info** from the **Administration panel**."

> **Tips or important notes**
> Appear like this.

Get in touch

Feedback from our readers is always welcome.

General feedback: If you have questions about any aspect of this book, email us at `customercare@packtpub.com` and mention the book title in the subject of your message.

Errata: Although we have taken every care to ensure the accuracy of our content, mistakes do happen. If you have found a mistake in this book, we would be grateful if you would report this to us. Please visit `www.packtpub.com/support/errata` and fill in the form.

Piracy: If you come across any illegal copies of our works in any form on the internet, we would be grateful if you would provide us with the location address or website name. Please contact us at copyright@packt.com with a link to the material.

If you are interested in becoming an author: If there is a topic that you have expertise in and you are interested in either writing or contributing to a book, please visit `authors.packtpub.com`.

Share your thoughts

If this book is helping you improve your skills, we'd strongly suggest leaving a review on `Amazon.com`. This helps us know whether you like our work and whether the chapter content has been valued, and also helps buyers on Amazon know whether the book is right for them.

So, everyone else benefits from your review, and we wouldn't want you to miss out. You can now reach out to `review@packt.com` with a screenshot of your review and the book URL, and we'll send you a $5 voucher for your next Packt purchase. Thank you in advance for engaging with us; we are excited to see your review!

Share Your Thoughts

Once you've read *API Analytics for Product Managers*, we'd love to hear your thoughts! Scan the QR code below to go straight to the Amazon review page for this book and share your feedback.

`https://packt.link/r/1-803-24765-7`

Your review is important to us and the tech community and will help us make sure we're delivering excellent quality content.

Download a free PDF copy of this book

Thanks for purchasing this book!

Do you like to read on the go but are unable to carry your print books everywhere? Is your eBook purchase not compatible with the device of your choice?

Don't worry, now with every Packt book you get a DRM-free PDF version of that book at no cost.

Read anywhere, any place, on any device. Search, copy, and paste code from your favorite technical books directly into your application.

The perks don't stop there, you can get exclusive access to discounts, newsletters, and great free content in your inbox daily

Follow these simple steps to get the benefits:

1. Scan the QR code or visit the link below

`https://packt.link/free-ebook/9781803247656`

2. Submit your proof of purchase
3. That's it! We'll send your free PDF and other benefits to your email directly

Part 1: The API Landscape

Before diving into the details of what API products are and how to build them, you will need to first look at the market landscape of API products to understand the products and services that exist in this space.

The history of the web-based APIs we know today can be traced back to the late 1990s when Salesforce launched a web-based sales automation tool. This application marks the beginning of the **Software as a Service** (**SaaS**) revolution. The World Wide Web has strengthened the underlying infrastructure that enables this newly discovered way of delivering software. Before the World Wide Web and the internet, APIs existed. Still, they were a form of proprietary protocol that supported small, distributed computer networks that spanned a limited area most of the time. The purpose of APIs in the pre-internet and post-internet eras is the same. APIs allow API providers to provide services, so external systems can call API providers to take advantage of those services. At the internet level, where developers worldwide create applications and host them on the World Wide Web, APIs are the future of the distributed computing paradigm.

We've come a long way since Roy Fielding's famous **REST dissertation**, which laid the groundwork for innovation in the API space. A few emerging technology businesses such as Salesforce, eBay, and Amazon paved the path for the current definition of web APIs. We've seen how Amazon Web Services laid the groundwork over the years while specific APIs, such as Flickr, have failed.

Once foundational APIs such as payment APIs, SMS, voice, and Google Maps started to come into being, the API economy had the foundation for a lot more to be built using these as building blocks. The 2007 release of the iPhone exponentially increased the speed of the API revolution and resulted in a vast universe of mobile apps that we know today.

In 2012, the then US president, Barack Obama, issued a comprehensive Digital Government Strategy aimed at delivering government data freely in machine-readable formats to enable researchers, innovators, and entrepreneurs to use and generate new products, services, and jobs. This has resulted in several APIs being launched by the US government over the years, such as airport delays, customer complaints, a Census API, `HealthData.gov` API, Healthcare Finder API, and Mars Weather API.

In 2014, tools such as Postman and GitHub became available, allowing developers to discover, evaluate, explore, and integrate with new APIs faster than ever before.

In 2015, NASA launched API.NASA.gov, where developers can learn to use existing NASA APIs or contribute their APIs to the catalog. In the following figure, you can view the visual timeline of some of the most noteworthy APIs and API tools that have been launched over the years that have shaped the landscape of the API business today.

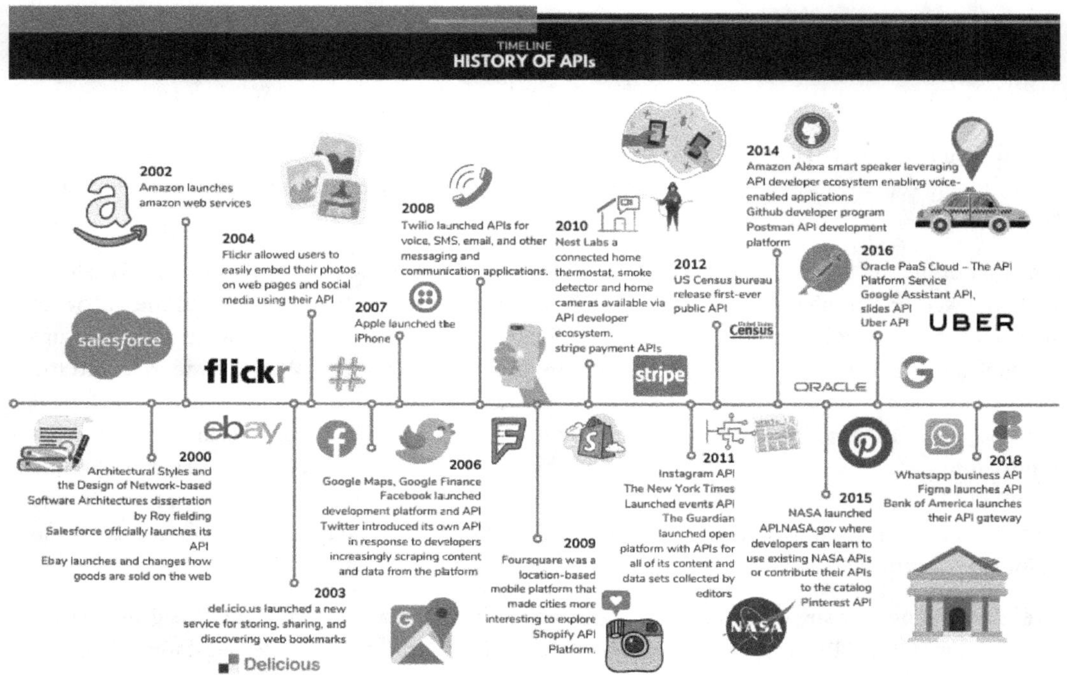

Figure P1.1 – Visual history of APIs

API products and services have evolved substantially in the last two decades. They are now at a point where various job functions have been established around APIs' development, maintenance, and support. The developer community is also a big part of the API landscape because developers actively explore new APIs to learn to integrate with them; this community has fueled an area of expertise around developer education such as developer evangelist and developer advocate functions.

Several aspects of API design guide us in building robust and scalable APIs. As multibillion-dollar companies get built on API-first business, an evolving function of API product management, measurement, and analytics for API products enables a methodical and systematic approach to creating successful APIs.

The pandemic effect

The outbreak of coronavirus acted as a forcing function that brought 332 million people online for the first time. This resulted in a digital transformation in shopping, socializing, and communicating. Voice and video APIs provided the infrastructure to build remote healthcare services that helped medical professionals serve their patients remotely. This also allowed children to attend school virtually and a vast population to work from home.

The pandemic fast-forwarded the plans of many businesses to come online by at least five years. Storefronts had to be closed overnight due to public health measures to avoid the spread of coronavirus; social distancing and stay-at-home rules catapulted heavyweights such as Walmart and Amazon to the front of the pack as consumers leaned on online shopping and grocery delivery more than ever before.

Platforms such as Etsy and Shopify allowed small businesses to set up e-commerce sites quickly. Shopify reported a record $2.4 billion in Black Friday sales globally in 2020. Etsy saw a 108% increase in sales from November 2019 to 2020, according to Edison Trends, a digital commerce research company.

APIs have been an integral part of this transformation, with voice and video APIs being the building blocks of all applications used across educators, healthcare, banking, and others. When a customer purchases something online, the transaction triggers a series of payment APIs to complete the transaction, email APIs to send confirmation, and ultimately, deliver shipment updates using package tracking APIs, SMS, and email APIs, to name a few among many APIs that are invoked for each transaction that a user makes.

API providers worked hard to scale their infrastructure considerably during the pandemic to deliver at such a scale. This has increased conversation around the maturity of APIs and usage monitoring and analytics of API products – topics we will be diving into in this book.

Great for business

SaaS has been the fastest-growing segment in the software revolution. According to IDG's 2018 Cloud Computing Survey, 73% of organizations have at least one application or a portion of their computing infrastructure already in the cloud. SaaS has dramatically lowered the total intrinsic cost of ownership for adopting software, solved scaling challenges, and removed the burden of local hardware issues.

APIs are a critical part of their strategy for fast-moving developers building globally. Instead of dedicating precious resources to recreating something in-house that's done better elsewhere, it is more time- and cost-effective to focus special developer efforts on creating a differentiated product.

For these and other reasons, APIs are a distinct subset of SaaS. By often exposing complex services as simplified code, API-first products are far more extensible, more accessible for customers to integrate into, and can foster a greater community around potential use cases.

At this point, several multibillion-dollar API-first companies are changing the way software is built.

Stripe is the largest independent API-first firm, with a market capitalization of over $95 billion in 2022. Stripe's early focus on the developer experience setting up and receiving payments helped it take off. It was even known as **/dev/payments** at first! Stripe's attention to creating idiomatic SDKs and beautiful documentation for each language platform allowed them to design the whole business around APIs.

Checkr is another excellent example of an API-first company simplifying the HR workflows of completing background checks on their employees and contractors, involving manual paperwork and the help of third-party services that spent days verifying an individual. This process had to be completed every time an employee had to get a new job or contract, which could be as frequent as every three months in the case of contract employees.

Checkr's API gives companies immediate access to various disparate verification sources and allows them to plug Checkr into their existing onboarding and HR workflows. It's used today by more than 10,000 businesses, including Uber, Instacart, Zenefits, and more.

Plaid delivers a similar service in the banking space by abstracting away banking relationships and complexities. Plaid started in 2013 and is currently valued at $13.4 billion as of 2022.

API tools such as Postman and Oracle Apiary have enabled the rapid evolution of the API economy and made working with APIs accessible to low-code/no-code customers by building UI-based tools.

This has given rise to an increasing number of jobs in software development, maintenance, and support of API products. Since APIs are significantly different from web- or mobile-based consumer products, and much of the technology and standards are still evolving, at this point, the most significant opportunity in software is to work on APIs and API-first companies.

This has given rise to an increasing number of jobs in software development, maintenance, and support of API products. Since APIs are significantly different from web- or mobile-based consumer products, and much of the technology and standards are still evolving, at this point, the most significant opportunity in software is to work on APIs and API-first companies.

Through the chapters of *Part 1*, you will learn about the application of product thinking for API products, which will enable you to build and grow API products. The following chapters will be covered in this part:

- *Chapter 1, API as a Product*
- *Chapter 2, API Product Management*
- *Chapter 3, API Life Cycle and Maturity*
- *Chapter 4, Building and Managing API Products*
- *Chapter 5, Growth for API Products*
- *Chapter 6, Support Models for API Products*

1

API as a Product

Most **application programming interfaces** (**APIs**) are sets of rules and protocols that allow different software applications to communicate with each other. They allow different software programs to interact with each other by exposing the functionality and data through a set of defined interfaces.

APIs are important because they allow different software programs to share data and functionality, which can greatly increase efficiency and reduce development time. They also allow for the integration of new technologies and services into existing systems, making it easier to add new features and capabilities.

APIs are becoming increasingly important in today's digital economy as they allow companies to share data and services with partners and third-party developers to create new products and services. They also allow for the automation of business processes and the creation of new revenue streams.

In this chapter, you will learn how to think about APIs as products. We will cover the following topics:

- Building with APIs
- Software-as-a-Service
- Establishing APIs as products
- Types of APIs
- Business models for API products
- Who builds APIs and who uses them?
- Notable API products that are shaping the API landscape
- Defining success for a product

By the end of this chapter, you will have learned about how you can think of APIs as products, how APIs are establishing themselves as a category of products, and some of the most prominent API products.

Building with APIs

APIs can help organizations become more efficient, make better use of data, and create new revenue streams. They can also open up new opportunities for innovation and provide a better customer experience.

Knowing how to work with APIs can be a valuable skill for developers, as more and more companies are looking for people who can integrate their systems with other technologies and services. Additionally, knowing how to use APIs can help developers create new applications that can take advantage of the data and functionality exposed by other systems.

APIs can provide several benefits to an organization, such as the following:

- **Increased efficiency**: APIs can automate business processes, allowing employees to focus on more important tasks
- **Improved data access**: APIs can make it easier to access data from different systems, which can be used for reporting, analysis, and decision-making
- **New revenue streams**: By making data and functionality available through an API, organizations can create new revenue streams by allowing third-party developers to access and use their data and services
- **Innovation**: APIs can open up new opportunities for innovation by allowing organizations to integrate with new technologies and services, and by enabling third-party developers to build new applications and services on top of an organization's data and functionality
- **Cost savings**: By exposing data and functionality through APIs, organizations can reduce the time and costs associated with developing and maintaining custom integrations between different systems
- **Better customer experience**: By making data and services available through an API, organizations can provide a more seamless and integrated experience for customers who use multiple channels to interact with them

With benefits such as increasing efficiency, improving data access, unlocking new revenue streams, and so on, APIs have become an exciting category of products. In the next section, you will learn about APIs as products and how they are used across industries such as e-commerce, finance, and so on.

Software-as-a-Service

Software-as-a-Service (**SaaS**) is a software delivery model in which a software application is hosted by a third-party provider and made available to customers over the internet.

Customers can access software and its functionality through a web browser or a mobile app without having to install or maintain it on their own servers. The SaaS provider is responsible for managing the infrastructure, security, and maintenance of the software.

SaaS is a type of cloud computing that enables customers to pay for the software on a subscription basis, usually on a monthly or annual basis. This allows companies of all sizes to access enterprise-level software without having to invest in expensive infrastructure and maintenance costs. Examples of SaaS include Salesforce, G Suite, Zoom, and Slack. SaaS is widely adopted in many industries, such as Customer Relationship Management (CRM), e-commerce, human resource management, project management, marketing automation, and many more. SaaS and APIs have a close relationship, as SaaS providers often use APIs to make their software available to customers.

APIs allow SaaS providers to expose the functionality of their software to external systems and applications. This allows customers to integrate the SaaS with other systems and automate workflows, such as integrating a SaaS CRM with a marketing automation tool or accounting SaaS with a website e-commerce platform.

APIs also allow SaaS providers to offer customization options to their customers, such as the ability to create custom reports or automate certain business processes. This allows customers to tailor the SaaS to their specific needs. With this understanding of the relationship between SaaS and APIs, you will learn to think about APIs as products in the next section.

Establishing APIs as products

When we say that an API is a product, it means that the API is being offered as a standalone service or offering that can be consumed by external customers or partners. In other words, the API is not just a means to an end but also a revenue-generating product in its own right.

APIs as products typically have their own pricing, service level agreements, and terms of service. They are often monetized through a variety of models, such as a subscription-based, pay-per-use, or revenue-sharing model.

APIs as products can be used to create new revenue streams for a company by allowing third-party developers to access and use their data and services. They also enable companies to access data, functionality, and services from other companies, enabling them to build new products, improve existing ones, and drive new business opportunities.

APIs as products can be used in a variety of industries, such as e-commerce, finance, healthcare, transportation, and more. Companies in these industries can leverage APIs to create new business models and disrupt traditional ones.

An example of how APIs allow companies to offer their products and services is how Uber uses the Google Maps API for mapping, uses PayPal APIs to offer a convenient way of making payments, and uses Twilio APIs to allow drivers and riders to communicate securely.

Uber is a ride-hailing service that allows users to request a ride through a mobile app. In order to provide its service, Uber uses a number of APIs, including Google Maps, PayPal, and Twilio:

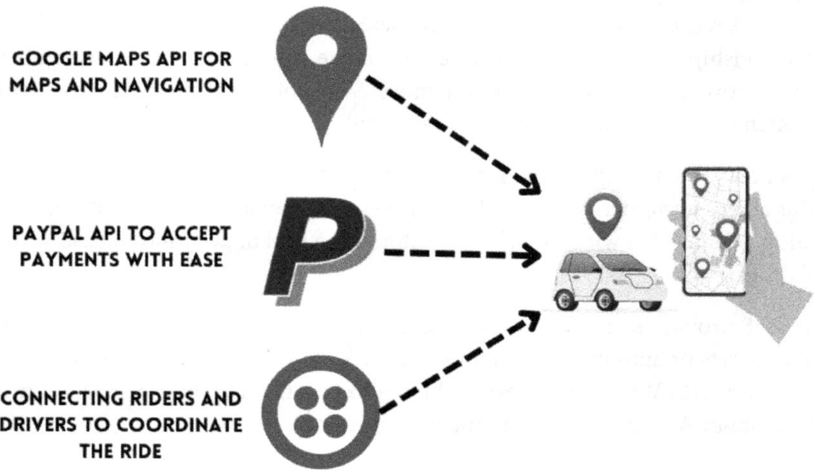
GOOGLE MAPS API FOR
MAPS AND NAVIGATION
PAYPAL API TO ACCEPT
PAYMENTS WITH EASE
CONNECTING RIDERS AND
DRIVERS TO COORDINATE
THE RIDE

Figure 1.1 – Examples of APIs being used during a single ride using a ride-sharing application such as Uber

The Google Maps API allows Uber to access the Google Maps platform and use its functionality within the Uber app. This includes features such as real-time traffic information, an estimated time of arrival, and turn-by-turn navigation. This allows Uber to provide accurate pickup and drop-off locations, an estimated time of arrival, and the best route for the driver to take to the rider's destination.

The PayPal API allows Uber to offer PayPal as a payment option to its users. By integrating with the PayPal API, Uber can securely process payments made through the app using the user's PayPal account. This allows riders to easily pay for their rides without having to enter credit card information.

By using these APIs, Uber is able to offer a more seamless and integrated user experience. The Google Maps API allows Uber to provide accurate and up-to-date information about pickup and drop-off locations, while the PayPal API allows for a convenient and secure way for users to pay for their rides.

Twilio is a cloud communication platform that allows businesses to programmatically make and receive phone calls and send and receive text messages using its APIs. Uber uses Twilio to provide a way for drivers and passengers to connect without revealing their phone numbers.

When a driver accepts a ride, the passenger's phone number is sent to the driver through the Twilio API, allowing the driver to call or text the passenger without ever seeing the passenger's phone number. Similarly, the driver's phone number is sent to the passenger through the Twilio API, allowing the passenger to contact the driver without ever seeing the driver's phone number.

By using Twilio's APIs, Uber is able to protect the privacy of its users by keeping phone numbers private. The Twilio API allows Uber to handle communication between drivers and passengers securely and efficiently. This way, it can enhance the user experience and help to improve the safety of the service.

To sum up, the use of the Twilio API allows Uber to use a cloud-based communication platform to connect drivers and passengers without revealing their phone numbers. This enables Uber to provide a more secure and efficient way of communication while protecting the privacy of its users.

Overall, the use of APIs such as Google Maps, PayPal, and Twilio allows Uber to access functionality and services provided by other companies, which they can then use to improve their own service and offer more features to their customers. This also saves them the cost of building any of these services by themselves while also reducing the time it takes to develop them.

Now that you have started to learn about how API products are making their mark, you will learn about different types of APIs in the next section.

Types of APIs

The Google Maps API is probably one that is most often used by people without realizing it because it is used via an interface, such as Uber or Lyft. APIs allow products to use capabilities from another product or company in a seamless way. This dramatically reduces the complexity of building software, as these capabilities are often so extensive that it is not possible to develop them from scratch.

There are three major types of API protocols and architectures:

- **Representational State Transfer** (**REST**): The most popular approach to building APIs is the REST architecture. REST is based on a client/server model and separates the frontend and backend of the API. This model allows for a great deal of flexibility in development and implementation. REST is *stateless*, which means that the API does not store any data or statuses between requests. For slow or non-time-sensitive APIs, REST supports caching, which stores responses. REST APIs, also known as *RESTful APIs*, can communicate directly or via intermediary systems, such as API gateways and load balancers.
- **Remote Procedural Call** (**RPC**): The RPC protocol is a straightforward way to send and receive multiple parameters and results. RPC APIs are used to perform actions or processes, while REST APIs are mostly used to share information or resources, such as documents. For coding, RPC can use two languages: JSON and XML; these APIs are known as JSON-RPC and XML-RPC, respectively.

- **Simple Object Access Protocol (SOAP)**: SOAP is a messaging standard defined by the World Wide Web Consortium and is widely used to create web APIs, typically with XML. SOAP supports many internet communication protocols, including HTTP, SMTP, and TCP. SOAP is also expandable and doesn't have a specific style. This means that developers can write SOAP APIs in different ways and quickly add new features and functions. The SOAP approach defines how the message is processed, including features and modules, the communication protocol(s), and the construction of the SOAP message.

Software architects will make the selection of the protocol depending on the use case that you are trying to serve with your APIs. There are various users and purposes for APIs, and you should be monitoring and managing them to verify that they are being used correctly. API products can fall into one of four categories:

- **Public APIs**: This is available for anybody to use. Good examples of public APIs are the APIs published by the US government, such as the Census API, which makes census data available to the public. The Google Books API also makes its entire database of books available via its public APIs. Public APIs may not always be free to use. Public APIs that are available for no cost are also referred to as **open APIs**.
- **Partner APIs**: APIs exposed by/to strategic business partners are known as partner APIs. They are not accessible to the general public and require specific authorization. While open APIs are entirely open, access to partner APIs requires an onboarding process that includes a particular authentication workflow.
- **Internal APIs**: Internal APIs, also known as **private APIs**, are accessible only through internal systems and are hidden from external users. Internal APIs are not intended for use outside of a company. They are limited to internal development teams to improve productivity and the reuse of services.
- **Composite APIs**: Multiple data or service APIs are combined to form composite APIs. They allow developers to make a single call to numerous endpoints. Composite APIs are useful in microservices architecture patterns where information from multiple services is required to complete a single task.

The type of API determines the user base that the API is targeted toward. You will need to identify and understand the unique needs of the audience and design the product in such a way that the customers are able to discover and use the right APIs for the desired use case.

Now that you have learned about the types of APIs, you will learn about various business models for APIs.

Business models for API products

In the context of APIs, the types of products across different business models are as follows:

- **Business-to-business** (**B2B**) APIs: These are APIs that are designed for use by other businesses. B2B APIs can provide access to a wide range of services, such as data analytics, financial services, and logistics management.
- **Business-to-consumer** (**B2C**) APIs: These are APIs that are designed for use by consumers. B2C APIs can provide access to services, such as weather forecasts, news updates, and social media platforms.
- **Business-to-business-to-consumer** (**B2B2C**) APIs: These are APIs that are designed for use by other businesses, but ultimately benefit consumers. An example of a B2B2C API would be an e-commerce platform API, which allows businesses to access inventory and customer data, but ultimately benefits consumers by providing them with a seamless shopping experience.
- **Consumer-to-business** (**C2B**) APIs: These are APIs that allow consumers to access and manipulate data and services provided by businesses. An example of a C2B API is a bank API that allows customers to check their account balances, view transaction history, and make payments.
- **Consumer-to-consumer** (**C2C**) APIs: These are APIs that allow consumers to access and manipulate data and services provided by other consumers. An example of a C2C API is a peer-to-peer marketplace API that allows users to buy and sell goods and services.

Overall, APIs can be used across different business models to provide access to data and services securely and efficiently and to enable new business opportunities and revenue streams.

As you begin to understand the types of APIs and the variety of business models for APIs, it is also important to understand the customers for APIs. In the next section, you will learn about API producers, API consumers, and the relationship between the two.

Who builds APIs and who uses them?

The entity that creates an API and makes it available for others to use is known as an **API producer**. The API producer is responsible for designing, building, and maintaining the API.

API consumer refers to the entity that uses or consumes the API provided by the API producer. The API consumer can be a developer, an organization, or another system that accesses the API to retrieve or update data or perform other operations. API consumers use the API created by API producers. You will learn more about the different life cycles of the API consumer and API producer in later chapters.

APIs are typically built by software developers who work for a company or organization that wants to expose certain functionality or data to other systems or applications. These developers create the rules and protocols that define how the API works, and they also create the code that implements the API.

APIs can be used by a wide range of people and organizations, depending on the purpose of the API. The customers for an API, also known as API consumers, can be broadly categorized into the following groups:

- **Internal developers**: These are the developers within the same organization that built the API, who use the API to access the data and functionality within the organization's systems. They may use the API to automate business processes, integrate systems, or access data for reporting and analysis.
- **External developers**: These are the developers outside of the organization who use the API to access the data and functionality provided by the organization. They may be third-party developers building applications that integrate with the organization's systems, or they may be partners or customers who access the organization's services through the API.
- **Business users**: These are the people within the organization who use the data and functionality exposed by the API to make decisions and run the business. They may use the data for reporting, analysis, and decision-making.
- **End users**: These are the users of the final product that uses the data and functionality exposed by the API.

APIs can be used by a wide range of people and organizations, depending on the purpose of the API. They can be used to automate business processes, integrate systems, access data, create new revenue streams, and improve the customer experience. APIs can also have different types of customers, such as developers, B2B customers, B2C customers, and so on, depending on the business model of the company providing the API.

The main goal of an API is to provide a way for different systems and applications to communicate and share data and functionality, and the customers of an API are the people and organizations that use that data and functionality to achieve their goals.

Now that you have developed an understanding of what API products are, their types, and the business models associated with them, we will take a look at some of the industry's most prominent API products in the next section.

Notable API products that are shaping the API landscape

Most ride-sharing companies use the Google Maps API in the background. E-commerce stores use APIs to update tracking information on orders and send shipping notifications to their customers. Services such as Shopify are built on a layer further abstracted where sellers don't have to make API integrations themselves but are offered the Shopify marketplace platform for e-commerce, which comes with nearly all the e-commerce-related integrations pre-built. Shopify integrates with PayPal using APIs, so the seller needs to provide their credentials for Shopify to connect with a PayPal account for their Shopify store.

These are some prominent API-first companies:

- Twilio
- Printful
- Twitter
- Tealium
- Plaid
- IMDB
- Amazon Selling Partner API
- Postman
- Marqeta

Let's take a look at each one of them and see how they position and support their products.

Twilio

Twilio is a cloud communications platform that enables developers to programmatically make and receive phone calls, send and receive text messages, and perform other communication functions using its web service APIs. Twilio has revolutionize authentication, **two-factor authentication** (**2FA**), as more users know what SMS and email codes are. Twilio has had a significant impact on the API industry by making it easy for developers to integrate communication functionality into their applications without having to build and maintain the underlying infrastructure. This has led to the creation of a wide variety of innovative communication-enabled apps and services. Additionally, Twilio's pay-as-you-go pricing model has made it accessible to small and large businesses alike.

Twilio uses web service APIs for programmable communication, such as making and receiving phone calls, sending and receiving text messages, and performing other communication operations. Twilio's API offerings across SMS, WhatsApp, voice, video, and email provide the building blocks to design highly customized and sophisticated customer interaction workflows. More than a million developers use Twilio, along with countless large brands.

Twilio's voice and video APIs were instrumental in enabling developers to build video applications that supported thousands of businesses that came online during the pandemic. The healthcare industry benefited the most by building secure voice and video applications to serve the community.

User verification is one of Twilio's most comprehensive services. Businesses can use SMS codes and programmable voice flows to verify user's identity.

Before Twilio disrupted contact center technology, most companies were using custom-built software that was hard to scale and maintain. Twilio is more advanced in contact centers compared to other domains in the same niche. It enables moving offline contact centers to the cloud quickly, all with built-in security, fraud prevention, 24/7 uptime, and so on.

Twilio is a favorite among the developer community, as it has one of the best-designed developer experiences. It has invested heavily in building a community that is innovative and engaged.

Printful

Printful is an innovative service that brings custom merchandise printing and embroidery to the masses. Printful has integrations with website-building platforms such as Shopify and Wix, where artists can upload their artwork to be printed on T-shirts, hats, socks, jackets, stickers, and more. The Printful APIs create and push them to Shopify. When a customer makes an order, Printful will automatically print and ship the product that the artist designed on its platform. Printful will also update shipping information passed on to Shopify via APIs. Shopify sends order tracking updates to customers using email templates. This integration makes it easy for thousands of artists to make money from their art, make it available all over the world, and handle large numbers of orders.

Previously, this kind of e-commerce platform was only available to large enterprises, and it took them many years, hundreds of employees, and millions of dollars to build.

Twitter API

The Twitter APIs have allowed several tools to be built on top of Twitter. Twitter APIs enable users to get Twitter data used to construct many sophisticated models, such as stock predictions based on Twitter sentiment analysis, and applications such as election predictions. The Twitter APIs also allow applications to post, retweet, and comment on tweets programmatically. By using these APIs, people have built many engaging Twitter bots, such as the Notion bot, which, when tagged by a user, saves the tweet to the user's Notion board automatically.

Twitter is also integrated with marketing management tools such as Hubspot, Canva, and so on.

Plaid

Banking and finance are heavily regulated spaces because of the sensitive nature of these organizations' data. As people start making more and more transactions online, their need to enable payments and connect various applications to banking has created the need for the abstraction of personal information while processing transactions.

Plaid is a FinTech company that establishes communication between applications, users, banks, and credit card providers. For many companies, it is impossible to integrate with the thousands of financial institutions that currently exist. Plaid has enabled these companies to simplify that process by acting as a middleman.

Plaid works without a standalone application and without the end user creating an account with Plaid. Plaid is integrated into other applications. Depending on the app's requirements, the service may appear as an option to add a bank account or link a different sort of account while you're using it. You'll be in Plaid's connection flow once you've been required to enter information, which usually comprises the following steps:

1. Choose your financial institution or bank.
2. Authenticate bank accounts by entering username and password information.
3. The authentication of the provided information takes place.
4. Choose which financial accounts you want to link.
5. The connection to the selected application or service is now complete.

Plaid verifies ownership of your bank accounts and captures the data points stated in the preceding section when you enter your username and password for those accounts. This data is shared with the application or service you're using. Services such as Venmo, Chime, Acorns, and Robinhood use Plaid to enable their users to connect their bank accounts with these services with ease.

Plaid provides various pricing options, starting with the free tier, which allows customers to try building and testing the core functionality such as transactions, Auth, balances, investments, and liabilities with up to 100 live items. The pay-as-you-go option offers unlimited items and no contractual minimums. Pay-as-you-go pricing provides customers with a way to provide pricing flexibility to customers. Avoiding contractual minimums allows more customers to try the product and grow their usage as they realize the value of the service. Most enterprises will have a more customized pricing option and work with a salesperson at Plaid to work through contracts.

Tealium

More digital experiences are powered by Tealium iQ™ tag management than any other corporate tag management supplier. Tealium iQ is the core of Tealium's customer data hub. It lets businesses manage and control their customer data and MarTech vendors across the web, mobile, IoT, and connected devices.

Tealium iQ is a feature-rich product with a unique tag management approach. Tags in Tealium iQ use a three-step template: tag configuration, load rules, and data mappings for adding and editing tags. The Tag Marketplace in Tealium is a pre-made tag template library of more than 1,000 tags ready for import. The user interface delivers data in rich dashboards for analytics and interpretation of data.

IMDB

IMDB is the most extensive online movie database, and over the years, it has been home to millions of movie reviews and ratings based on user-generated data. It also catalogs all the cast and crew of movies, TV shows, and more. This rich data can be accessed using IMDB APIs, available only in a paid model with prior approval.

Amazon Selling Partner API

The Amazon Selling Partner API (SP-API) is a service that Amazon offers to developers that lets them use all of the old features of Amazon Market Web Services. This API is updated with features such as a REST-based format and JSON outputs. The API allows you to create applications for your personal Amazon seller account, applications for sellers to authorize and use to help run their Amazon businesses, and applications to be published to the Amazon Marketplace Appstore.

One example of an application built using the Amazon API is `camelcamelcamel.com`, which tracks and monitors prices for any product specified by a user and notifies the user when the prices drop to the desired value. This service also monitors the prices of products and shows the highest and lowest prices over time.

Postman

Postman is the standard API development tool in the industry. It is used to build, test, catalog, and change APIs. Postman allows users to use APIs in a user-friendly GUI and eliminates the need to write code to work with APIs. Postman also lets users add predefined API calls to a collection that other users can import and share. Postman also supports access to popular web API clients. The AWS documentation for Postman is well laid out and provides a step-by-step guide to creating an API project with it.

The Postman API allows users to access data stored in the Postman account programmatically.

Postman has also built an amazing community of developers through its platform, blogs, and developer evangelist videos, which interact with the developer community and make APIs easy to use.

Marqeta

Marqeta is another notable FinTech API start-up that executed its **initial public offering** (**IPO**) in 2021. It is a cloud-based, open API platform that allows consumers to create customized cards. Businesses may use Marqeta to provide their consumers with payment cards without dealing with banks. Marqeta manages the payment technology's backend, lowering the integration cost for businesses and allowing them to issue cards to their consumers swiftly.

On-demand delivery services such as DoorDash and Uber give new employees payment cards powered by Marqeta that can be used at stores and restaurants to keep up with the rise in orders. Using Marqeta's APIs, they are able to onboard new workers quickly and use Marqeta's card servicing to provide their users with a standard experience in terms of being able to report transaction disputes, issue new cards, or report stolen cards.

Marqeta also provides *buy now, pay later* products that enable services such as Affirm and Klarna to pay their merchants. Corporate credit cards are also starting to be powered by Marqeta APIs for a smooth employee experience. Large financial institutions, such as JP Morgan and Goldman Sachs Marcus, have begun to use Marqeta to issue virtual cards to their customers because of the easy API integration.

Meanwhile, developers at crypto exchanges such as Coinbase and ShakePay use Marqeta's APIs to allow customers to convert cryptocurrencies into government-backed fiat currency at the point of sale. Businesses can build their applications using Marqeta's APIs, where users can sign up for cards. Marqeta has a pricing model that is based on how much you use it and offers complex dispute flows that lower the costs of running cards.

Defining success for a product

The product manager is in charge of setting up metrics for the whole customer journey and connecting them to the business process that goes with them. Product managers usually work with data analytics teams to instrument the data needed to deliver the necessary analytics to measure product adoption, usage, and retention.

Whether it's a SaaS product, a physical product, an IoT product, or an API product, a company's objective is to monetize the product in a way that benefits the company while also providing value to customers. Customer value can be measured in many ways, such as by increasing sales, increasing net margins, lowering operational costs, and keeping customers from leaving.

There is always a need to decrease customer turnover, regardless of the product or service. Product owners are responsible for producing value for their goods and controlling customer churn, whether they are SaaS products such as Slack, Dropbox, or Coupa, or tangible items such as smartphones or video game consoles.

Customer centricity begins at the very beginning of the customer journey: marketing. When the messaging for a product is clear regarding the value it delivers to the customer, the right customers are drawn to the funnel. A good customer experience will come from an optimized funnel, a smooth onboarding process, and good customer service.

As you have learned in this chapter, there are several ways that different companies design their product offerings to drive value and ease for their customers. Putting customers at the center of your product thinking allows you to design products that are tailored to your customers' specific needs.

Summary

At this point, you should be familiar with the various ways that products are categorized. Understanding the types of products will help you analyze the unique challenges of any product based on its technology, business model, and the relationship between API consumers and API producers.

Looking at API products, you can now understand the architectures and protocols used and the types of APIs across private, public, partner, and internal APIs. With this understanding, you can look at the wide variety of APIs currently on the market and understand how these protocols enable the development of unique API-first products and business opportunities. You are now ready to think of APIs as products.

In the next chapter, you will learn about API product management and develop product thinking to enable you to build an API-first strategy for your organization.

2

API Product Management

Product management is a relatively new discipline, but it has taken the industry by storm. There are over 280,000 product management jobs posted just on LinkedIn as of the writing of this book, and the average salary of a product manager is $113,446 per year as of 2022. But what is product management? What do product managers do? How do you become a product manager?

In this chapter, we will dive into the role of a product manager and look at the various functions that this role collaborates with, along with the types of product management roles and specializations that exist. We will learn about some foundational concepts in product management that all product managers must know and then dive into how to apply these concepts to API product management.

We will cover the following main concepts:

- The role of the product manager
- Types of product managers
- Responsibilities of an API product manager

The role of the product manager

Product management is the process of planning, developing, launching, and marketing a product. It involves identifying customer needs, developing a product strategy, and overseeing the development and delivery of a product or service to the market.

Product management is a cross-functional role that involves working with different departments such as engineering, design, marketing, and sales. The product manager is responsible for defining the product vision and strategy, setting product goals and objectives, and ensuring that the product meets the needs of the target market.

The main responsibilities of a product manager include the following:

- **Identifying customer needs and market trends**: Conducting research to understand customer needs and market trends, and using that information to define the product vision and strategy
- **Developing a product roadmap**: Creating a plan for the development and launch of a product, and ensuring that it aligns with the overall company strategy
- **Managing the product development process**: Overseeing the development of the product, including working with engineering, design, and other teams to ensure that the product is delivered on time and within budget
- **Defining product features and requirements**: Identifying the features and requirements that the product must have to meet customer needs and achieve the business goals
- **Creating product marketing and launch plans**: Developing plans for how to market and launch the product, and working with the marketing and sales teams to execute those plans
- **Monitoring and analyzing product performance**: Tracking the performance of the product after launch, and using that information to make decisions about future product development and enhancements

Product management is a critical role that helps organizations develop, launch, and market products that meet the needs of customers and achieve business goals. Therefore, developing a systematic understanding of customers' goals and pain points is the most critical focus for product managers. Be it building a new product (from zero to one) or evolving an existing product, you will need to use all the resources available to be able to develop a deep understanding of the customers so that you can set product strategies and roadmaps that drive success.

Product management used to be part of the marketing or engineering departments, which resulted in prioritization and focus conflicts. Product management is now a standalone function and reports directly to the **Chief Executive Officer** (**CEO**) or **Chief Product Officer** (**CPO**). This aligns the product team with the company's vision and goals, makes them both internal and external evangelists for the vision, and gives them the autonomy to make difficult prioritization decisions.

An example of a company that has made product management a standalone function and reports directly to the CEO or CPO is Slack. Slack is a popular communication platform. It was developed by the company Tiny Speck but later rebranded as Slack Technologies.

In the early days of the company, product management was part of the engineering department, and decisions were made by engineers. As the company grew, the product team realized that this structure was not working, as the product team's goals and priorities did not align with those of the engineering department.

To address this issue, the company decided to make product management a standalone function and have the product team report directly to the CEO or CPO. This allowed the product team to focus on the company's overall vision and goals and make difficult prioritization decisions without being influenced by other departments.

As a result, the product team was able to develop a product that met the needs of its customers and helped the company achieve its goals. The company's success with this approach has been instrumental in making Slack one of the most successful communication platforms on the market.

This example shows how aligning the product team with the company's vision and goals and giving them the autonomy to make difficult prioritization decisions can lead to the development of a successful product and business. As you have seen in this example, product managers work cross-functionally with various teams to accomplish their goals.

Since product managers have such a vast set of responsibilities, they need a variety of skills to be successful in their roles. You will learn about the skills that allow product managers to partner with various teams and stakeholders in the following sections.

Product thinking

Product thinking is the ability to identify and solve problems that customers have in order to create a product that meets their needs. Product sense is the ability to understand what makes a product successful and to make decisions that will improve it.

Product managers can develop these skills by gaining a deep understanding of their customers and the market, constantly testing and iterating on their products, and learning from the successes and failures of other products. Additionally, staying up to date with industry trends and best practices can also help product managers develop their product thinking and sense.

An example of product thinking for an API product would be identifying a problem that developers face when using existing APIs, such as a lack of documentation or poor performance. The product manager would then come up with a solution, such as creating an API that has detailed documentation and is optimized for speed.

An example of product sense for an API would be understanding that developers are looking for an API that is easy to use and has a wide range of functionality. The product manager would then make decisions that would improve the API, such as adding more endpoints and simplifying the authentication process.

Additionally, the product manager should have a clear understanding of the target market, the competition, and industry trends to make sure the API is fulfilling market needs in the best possible way. In order to understand customers, and develop and launch products, product managers would partner with various stakeholders, which you will learn about in the next section.

Stakeholder management

You will often hear that being a product manager requires you to develop people skills. By now, you have probably realized the various teams that product managers interact with and work with on a regular basis. This would inevitably require you to build empathy, not just for the end customer but also for all the team members and stakeholders involved in shipping the products.

The following diagram shows the variety of partner teams and stakeholders that product managers work with at all times to drive product success:

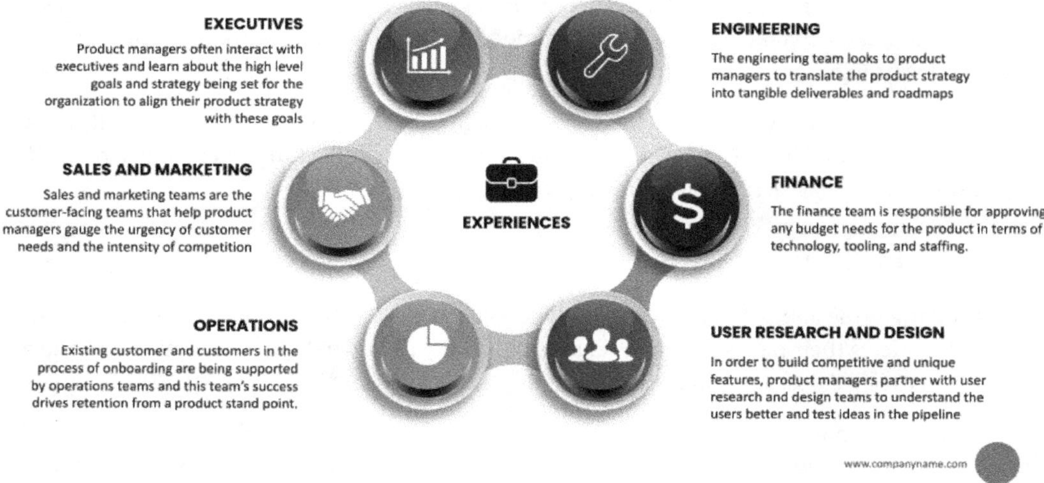

EXECUTIVES
Product managers often interact with executives and learn about the high level goals and strategy being set for the organization to align their product strategy with these goals
ENGINEERING
The engineering team looks to product managers to translate the product strategy into tangible deliverables and roadmaps
SALES AND MARKETING
Sales and marketing teams are the customer-facing teams that help product managers gauge the urgency of customer needs and the intensity of competition
EXPERIENCES
FINANCE
The finance team is responsible for approving any budget needs for the product in terms of technology, tooling, and staffing.
OPERATIONS
Existing customer and customers in the process of onboarding are being supported by operations teams and this team's success drives retention from a product stand point.
USER RESEARCH AND DESIGN
In order to build competitive and unique features, product managers partner with user research and design teams to understand the users better and test ideas in the pipeline
www.companyname.com

Figure 2.1 – Stakeholders for product managers

In addition to research, design, development, analytics, and support, you will also often partner with other product managers across the organization. Products within the same organization can have overlapping or complementary functionality that needs to be developed collaboratively. You will need to build awareness of not just your own product but also other products across the company to look for these opportunities to team up to expand the impact of the features your team works on as well as ensure there are no redundancies in the work being done.

In order to be able to drive product goals across various stakeholders, you need to develop a deep understanding of the product life cycle, which you will learn about in the following section.

Understanding the product life cycle

Whether you are building a new product from scratch or iterating upon an existing product, whether you are working for your own start-up or for a corporation, the process of building products starts with understanding the product life cycle. The product life cycle is the highest level of product thinking that will determine the big-picture goals and set the strategy for all the work you do as a product manager for a given product.

Conceptually, a product has five stages in its life cycle, as we can see in the following figure:

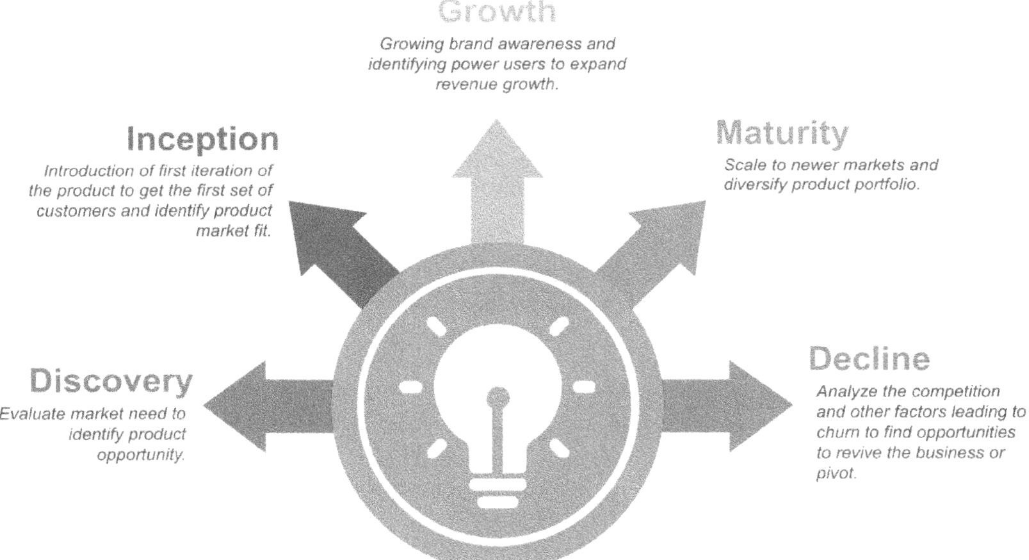

Figure 2.2 – Five stages of the product life cycle

Let's go into the details of these five stages:

- **Discovery**: Each product begins with a problem that needs solving and a customer need that is unmet by the current products on the market. As a founder or a founding product manager, the first step to do is to study the market in terms of customer segments and competitive products and identify the opportunity for building a product that can serve a customer's needs. This would form the foundation for creating a product strategy and building a business plan.

- **Inception**: Once you have a product in development, it is time to start finding the first set of customers. In this phase, the key goal is to raise awareness of the product and develop an audience for the product. This is a very marketing - and sales-driven phase for the product, and as the first wave of customers comes on board, this is the foundation of product features and the customer base for the product. If, at this stage, the customers are pleased with the product, the next phase of growth will become much more efficient.
- **Growth**: If the product becomes successful, it advances to the growth stage. As the product has found a product-market fit and this stage is marked by increased demand, increased production, and increased availability, you will focus on scaling the infrastructure and opportunities to make sure the customer experience doesn't deteriorate with the increasing customer base.
- **Maturity**: As production and marketing costs decrease, this is the most profitable stage. The infrastructure is able to sustain the growth and the product is at a stage where the most refinement happens. At this stage, the challenges become more specific and complicated. At a mature stage, you have enough customers to significantly focus on the existing user base for research purposes as it is going to be a significant sample size.
- **Decline**: Once a product starts to succeed, multiple players often start to create similar products to compete for the market share. With increased competition, competitors might be able to offer better features or lower prices. The product's market share may begin to dwindle and decline. It can be quite a challenge to try and revive a product that is on the decline, but it is not impossible.

As new products begin to be successful, they grow in demand and popularity. This popularity of new products overwhelms the old ones and effectively replaces them. As new products grow, companies tend to curtail marketing activities. This is because it reduces the cost of manufacturing and marketing the product. If the demand for a product drops, it can be withdrawn entirely from the market.

Based on where a product is in its life cycle, you can determine the goals and challenges for the product. This would also help you determine the right teams to form or partner with as you start working on product initiatives.

During the discovery phase, if you are building a brand new product, market research would drive the decision-making for setting the product strategy. Once you understand that market opportunity, you can identify whether there is a need and an audience for what you are trying to build. In the early stages of product development, there are various research methodologies to test the product using low-fidelity mockups to validate product ideas without investing development effort, which can be expensive and time-consuming.

When you work on an existing product, the market and user research doesn't stop but continues to guide the product strategy in terms of how the product evolves and grows to address more use cases and wider audiences. The market research efforts for an existing product are focused on identifying the next market opportunity for the product.

Understanding the product life cycle will help you determine the high-level strategy for the product and be able to identify high-level opportunities to make an impact on product success. If the high-level strategy for the product is well aligned with the success of the business, your product decisions will make more sense to stakeholders in marketing, customer success, engineering, and executive management, and you will be able to establish better partnerships across all these teams.

These concepts can apply to all types of products across all industries, and learning these skills will help you not just become a more effective product leader but also an entrepreneur if you ever desire to do so. Product managers are often responsible for setting high-level strategies for their products, and a deep understanding of the phase in which the product is in its product life cycle can help you identify the right priorities for product initiatives. Next, we will learn about how market research also fuels product strategy.

Market research

Market research is critical not just for early-stage products; it is also an activity that, ideally, should never stop and should continue to inform the product strategy. Market research is the practice of gathering information about your target market and helping to validate the success of a new product, iterating on an existing product, or how people feel about your brand to make sure your team is effectively communicating the value of your company.

Market research can provide answers to a variety of questions about the state of an industry, but it is far from a crystal ball on which you can rely for customer insights. Market researchers look into various aspects of the market, and it can take weeks or even months to develop an accurate picture of the business landscape.

Market research provides insights into a wide range of factors that can inform product strategy, such as the following:

- How and where do your target audience and current customers go to research products or services?
- Which of your competitors does your target audience seek information, options, or purchases from?
- What are the pricing attitudes for a specific product or service?
- What are the popular products in your industry and among your buyers?
- Who are your competitors and what are their challenges?
- What factors influence your target audience's purchases and conversions?
- Consumer perceptions of a specific topic, ailment, product, or brand.
- Whether there is demand for the business initiatives in which you are investing. Unaddressed or underserved customer needs can be converted into a product opportunity.

Market research can be a combination of qualitative and quantitative data, which is crucial for you to be able to create a competitive product strategy that you can present to executives and get buy-in for.

User research

Product managers work with user research teams to develop qualitative insights and user personas. User research is more specific than market research and focuses on developing a deeper understanding of the target audience for your product. User research complements the quantitative data from user analytics with qualitative insights about the customer population. Together, this helps build an in-depth knowledge of the user personas that the product targets. Added to this, you will need to have a keen interest in competing products, determine the role of the product in the bigger scheme of a business, and define how it will be marketed to customers.

User research is focused on answering some of these key questions about the customers:

- What do users need?
- What are the most important customer pain points?
- How can you solve customer pain points to drive value?

During the product discovery phase, you can focus user research efforts on user interviews and contextual inquiries. As the product enters the testing phase, you want to switch focus to dogfooding, concept testing, and usability testing:

- **Dogfooding** is the practice of having the team use the product to gain the user's perspective on the experience and be able to find gaps and opportunities for improvement.
- **Concept testing** allows product managers to find a way to test the concept of the product without building the product; this could be wireframing or creating high-fidelity mockups of the product.
- **Usability testing** involves seeing actual users interact with a website, app, or other digital product to evaluate how well it functions. Typically, researchers who work for a corporation are the ones who observe the users. Usability testing aims to identify areas of uncertainty and find possibilities to enhance the user experience.

Once a feature has been launched, user research efforts are focused on conducting surveys to find ways to improve the experience, as well as measure customer satisfaction with the product and listen to customer sentiment.

Experimentation and hypothesis testing

It is preferable to focus on incremental improvements that benefit the end user as they are implemented. This allows you to avoid making assumptions about what users want and the best solutions; you can test each assumption and hypothesis by performing a test that isolates the effect of each change. Products such as Facebook, Twitter, and Amazon have hundreds of A/B tests on any given day.

This approach enables a product team to get real-time user feedback and then adjust their strategy accordingly. Making and testing hypotheses is a less expensive and easier way to iteratively improve product value while also allowing for course corrections.

Hypotheses can be tested using a variety of product hypothesis testing methods to ensure that users continue to receive value from a product. To test a hypothesis, as outlined previously, a corresponding test design is necessary. For example, new features and changes can be measured in terms of their value, the most valuable features can be highlighted, and incremental improvements can be made.

The following are the different types of hypothesis testing:

- **Product A/B testing**: Randomized A/B testing validates the most common use cases by releasing a change or feature to half of the users and withholding it from the other half. Half of the users will see the change and half will see the old website. Each group's conversion will be measured and compared. If the group shown larger product images converts more, the original hypothesis was correct, and the change can be made for all users.
- **Multivariate testing**: Sequential testing can be slow, especially with multiple versions. Instead of testing sequentially, a multivariate test can be used, in which users are split into multiple variants. For example, four groups (A, B, C, and D) each have 25% of users; group A users won't see any changes, but group B, C, and D users will. Multiple variants are tested against the current product version to determine the best variant.
- **Testing before/after**: Sometimes, network effects prevent splitting users in half (or into multiple variants). If the test involves determining which logic for calculating Uber surge prices is better, the drivers cannot be divided into different variants because the logic considers the city's demand-and-supply mismatch. In such cases, a test must compare before-and-after effects to reach a conclusion. The problem is that seasonality and externalities can affect the test and control periods differently. At some point in time, Uber's surge pricing logic changes from A to B. Comparing the effects before and after that time doesn't prove that the logic change caused them. Different demands or other factors may have caused the difference resulting in a limitation of your ability to compare before-and-after results.
- **Hypothesis testing using timed on/off**: Time-based on/off testing can overcome the downsides of before/after testing by introducing the change for a certain period of time, turning it off for an equal period of time, and repeating it for a longer duration. This method reduces the effects of seasonality and externality, making tests more robust.

Agile methodology

Product managers are closely embedded within an engineering team that works on the product and features that the product manager prioritizes for development. Engineering teams typically report to engineering leadership and are designed in such a way that, within the team, there are all the technical skills required to build the product.

Agile is simply one method of product planning and development. While there are numerous approaches, processes, and technologies that teams can use to apply agile concepts to their work, prioritizing customer needs and creating an amazing product is the most important thing.

Most commonly, agile teams work in 2-week sprints but this can vary based on the engineering practices of the organization. Agile establishes how all of these roles work together over the course of the sprint duration to set the goals for each sprint, evaluate the outcomes of the last sprint, and prepare the requirements for the next sprint. The agile methodology defines four key agile roles:

- **Development team**: An agile development team is a group of people with skills such as design, development, testing, and delivery.
- **Owner of the product**: A product owner's job is to make sure that the product made by the development team is as valuable as possible. This role collects technical requirements, works on the product backlog, and writes up user stories.
- **Scrum master**: The scrum master leads the team through the agile process so that the work that the product owner thinks is most important can be done, and makes sure the development team receives the necessary inputs and feedback from the product owner at various stages of development. The scrum master also establishes and hosts operational meetings and sprint rituals, such as backlog grooming, sprint planning, sprint demos, and sprint retrospectives.
- **Stakeholders**: Stakeholders can be anyone who has a stake in how a software project turns out. This includes a wide range of people, such as end users, executives, IT, operations, portfolio managers, and support.

Some of the key operational concepts and sprint rituals that are part of the agile process that you will follow are listed here:

- **Sprints**: A sprint is a brief period of time during which a development team works to finish particular assignments, benchmarks, or deliverables. Sprints, also known as **iterations**, essentially divide the project schedule into manageable time blocks in which smaller objectives can be achieved. Two-week sprints are most common and, as a best practice, sprints are usually not longer than 1 month.
- **Backlog**: A product backlog is a list of new features, enhancements to existing features, fixes for bugs, changes to the infrastructure, and other tasks that a team may carry out to accomplish a certain goal. The only reliable source for the items a team works on is the product backlog.

- **Sprint backlog**: In order to complete the sprint goal and advance toward a desired end, a team targets delivering a subset of the product backlog known as a sprint backlog.
- **User stories**: The product owner breaks the work into functional units (known as user stories) from inputs from customer research. One of the most frequently suggested tools for writing user stories is the user narrative template: "`As a {customer persona},I want {to take an action} So that {I can have a desired result}`."
- **Backlog grooming**: Backlog grooming is the process of the product owner and some, or all, of the other team members regularly revising the backlog to make sure it contains the right things, that they are prioritized, and that the items at the top are prepared for being worked in on the upcoming sprints.
- **Sprint planning**: The team decides which items from the product backlog they will work on during the sprint through a process called **sprint planning**, which takes place at the start of a sprint.
- **Sprint demo**: At the end of each sprint, the team meets to review all the work that was done and demonstrate the new features that were developed as part of the sprint.
- **Sprint retrospective**: To make sure there is an opportunity to learn from each sprint, teams retrospect on all the things that went well and all the things that didn't as part of the completed sprint. This allows team members to look back at the processes and outcomes and suggest improvements.

You will partner with your engineering manager or a program manager to run the sprint rituals across your engineering team. It is the responsibility of the product managers to make sure the backlog is being populated and prioritized so that the team has a view of upcoming tasks at all times. As new features are being developed and bugs are being fixed, you will be able to see the impact of the work of each sprint in product analytics. You will learn more about data analytics in the following section.

Data analytics

Over the past 3 decades, technology has enabled an enormous amount of data to be generated, collected, and stored. We can now make data-driven decisions for things we could never do in the past. Market research and user research tools have made experimentation and feedback gathering possible at scale. With this availability of data, the job of a product manager has also expanded to include making product design and development decisions based on data.

Product managers should have a good understanding of data and analytics and know how to use tools such as Excel and Google Analytics to track product performance. Additionally, many product managers also use **Structured Query Language** (**SQL**) to work with data stored in databases. SQL is a programming language that allows you to access, manipulate, and analyze data stored in relational databases.

Other common tools that product managers use include the following:

- **A/B testing** tools, such as Optimizely, Google Optimize, and VWO, to test different product features and measure their impact on user behavior
- **Business intelligence** (**BI**) tools, such as Tableau and Looker, to visualize and analyze large amounts of data, and create interactive dashboards and reports
- **Heat mapping** tools, such as Hotjar and Crazy Egg, to track user interactions and understand how users navigate through a website or app
- **Survey** tools, such as SurveyMonkey and Typeform, to gather feedback from users and understand their needs and preferences
- **Mixpanel, Amplitude, Heap, and other analytics platforms**, to track user behavior, measure engagement, and track key metrics
- **Excel**, to organize and analyze data, create pivot tables and charts, and perform basic statistical analysis

It's important to note that the specific tools a product manager uses will depend on the company, the product, and the specific data needs. However, having a good understanding of these tools and being able to use them effectively can help a product manager make data-driven decisions and improve the product.

There are two key scenarios where you will use analytics most commonly:

- **Establishing metrics or key performance indicators** (**KPIs**): When building products, you will need to build a hypothesis around the impact the development of a feature or product will make. This could be in terms of improving customer experience, acquisition, or retention, or reducing churn. As you build and ship the product or feature, you will closely monitor the set metric and see how the metric is impacted and whether the changes you made are having the desired impact on the metric.
- **Diagnosing KPIs**: Product managers are often faced with scenarios where there is a change in a metric, and you have to diagnose the metric to identify what could be causing this change. This could be a drop in visitors to a web page caused by a natural calamity in a certain part of the world, or it could be because of a buggy feature impacting a specific segment of customers. In large organizations where many teams are shipping products and features simultaneously, metrics can also be impacted by another team's efforts inadvertently, and you will need to diagnose the impact on the metrics you track back to the responsible team to make decisions on actions that must be taken to improve the metric.

Most of the time, you will work with a data analyst to accomplish your data analysis needs, but in small organizations, sometimes product managers will simply learn to use tools such as Looker and SQL to self-serve their data analysis needs.

Customer feedback channels

The best way to develop customer empathy is to interact with customers. Customers who interact with customer feedback channels (such as chat, phone, or support tickets) essentially create a wealth of data that is very insightful for product managers to use to understand the customer experience.

With early-stage products, you have the luxury of working more closely with the initial customers and learning about their experience with the product first. As the customer base grows, it's not possible for product managers to engage with customers, and a customer support team might handle incoming customer requests.

Product managers partner closely with customer support teams to gain insights into the most common issues customers face and use this information to guide the product roadmap. Customer support teams must also be informed of upcoming customers and product features so that they can anticipate customer questions once those features are launched. This allows for a healthy exchange of insights, which helps both support teams and product managers serve the customer best.

In the case of APIs or other enterprise-facing products, sales and solution architecture teams might also be involved with customers. Product managers can connect with these teams to get in touch with the customer and learn about their experiences, needs, and feedback.

Types of product management roles

In the previous section, we learned about all the different stakeholders that product managers work with, but there are certain specializations that product managers also bring to the team. For example, a product manager with a background in marketing or sales could unlock growth strategies for the product and might be well suited to work with products that are in the growth life cycle.

There are many different types of product manager roles, and the specific responsibilities vary from company to company. However, some common product manager roles include the following:

- **Growth product manager**: Responsible for driving user acquisition and engagement for a product, they often focus on metrics such as user retention and lifetime value
- **Data product manager**: Responsible for creating products that leverage data to provide insights or drive business decisions
- **Platform product manager**: Responsible for creating a platform that allows other products or services to be built on top of it
- **API product manager**: Responsible for creating and maintaining an API that allows external developers to access the functionality of a product or service
- **Business product manager**: Responsible for developing a product that meets the needs of a specific business unit or department, and often works closely with sales and marketing teams

- **Technical product manager**: Responsible for the technical direction and execution of a product, and often has a background in software engineering or a related field
- **B2B product manager**: Responsible for developing products for businesses rather than consumers, and may have a different approach to user research, pricing, and distribution than B2C product managers
- **B2C product manager**: Responsible for developing products for consumers, and typically focuses on user experience and design, as well as pricing and distribution strategies

These are just a few examples of the different types of product manager roles that exist. The exact responsibilities and focus of the role can vary depending on the company, industry, and stage of the product. In the next section, we will dive deeper into the roles and responsibilities of the API product manager.

The responsibilities of an API product manager

Added to the understanding of product management principles, the management of an API product also requires knowledge of various types of APIs, the API life cycle, maturity, support models, and so on. The B2B nature of APIs presents challenges in terms of user research, testing, operation, and maintenance. APIs also have a wide range of stakeholders who are deeply invested throughout the product life cycle.

Product management has become increasingly important for APIs as the use of APIs in the digital economy has grown. This is because APIs are becoming a critical component of many products and services, and managing them effectively is crucial for the success of these products and services.

The key responsibilities of API product management include the following:

- **Defining the API strategy**: Product managers are responsible for defining the API strategy and ensuring that it aligns with the overall product and business strategy. This includes identifying the target market, determining the API's key features and functionalities, and creating a plan for API development, launch, and maintenance.
- **Managing API development**: Product managers are responsible for overseeing the development of the API, including working with engineering and design teams to ensure that the API meets customer needs and is delivered on time and within budget.
- **Creating API documentation and developer portals**: Product managers are responsible for creating API documentation and developer portals that make it easy for developers to understand and use the API. This includes creating code examples, tutorials, and other resources that help developers integrate the API into their products and services.
- **Identifying and analyzing API use cases**: Product managers are responsible for identifying use cases for the API and analyzing how customers are using the API to ensure that it is meeting customer needs and achieving business goals.

- **Managing API monetization**: Product managers are responsible for identifying and implementing monetization strategies for the API, such as usage-based pricing or tiered access levels.
- **Monitoring and analyzing API performance**: Product managers are responsible for monitoring and analyzing the performance of the API after its launch and using that information to make decisions about future API development and enhancements.

Overall, product management has become increasingly important for APIs as they have become a critical component of many products and services, and managing them effectively is crucial for the success of these products and services.

An API product manager requires foundational product management skills as well as a deep understanding of API-specific concepts, such as the API development life cycle, governance, maturity, and developer experience, which would allow them to develop long- and short-term strategies for API products. This is a big part of the responsibilities of an API product manager.

Useful terminology for API product managers

APIs have really evolved into a specialized product category because of the uniqueness of how they are developed, introduced, and iterated over time, as well as the uniqueness of the skill set that is required. As an API product manager, you will be responsible for leading the product through the entire product life cycle, from discovery through maturity, growth, and, occasionally, customer interaction. The key customers for APIs are developers, and developers are a unique audience compared to other market segments, which presents its own set of challenges for the API producer. In this section, you will learn about some of the key concepts and terminology that every API product manager must know.

As an API product manager, you will often talk to developers, both on your team and as customers. You must have a strong understanding of various API terminology to effectively manage and market your products. Some of the key API terminology that you should be familiar with includes the following:

- **API endpoints**: The specific URL or location that an API request is sent to. Each endpoint corresponds to a specific function or service provided by the API.
- **API resources**: The data or functionality provided by the API, typically represented as a collection of endpoints.
- **API methods**: The actions that can be performed on an API resource, such as reading, writing, or updating data. Common methods include `GET`, `POST`, `PUT`, and `DELETE`.
- **API authentication**: The process of verifying the identity of the API user, usually through the use of an API key or OAuth token.
- **API authorization**: The process of determining whether an authenticated user has permission to access a specific resource or perform a specific action.
- **API rate limiting**: The process of limiting the number of requests that can be made to an API within a given time period to prevent overuse or abuse.

- **API versioning**: The process of creating and managing multiple versions of an API to support backward compatibility and allow for updates without breaking existing integrations.
- **API gateway**: A server that acts as an intermediary between an application and a set of microservices. It handles tasks such as authentication, rate limiting, and request routing.
- **API management**: The process of designing, publishing, documenting, and maintaining APIs. It includes functions such as security, analytics, and developer portal.
- **SDK**: The **Software Development Kit**, a set of tools and libraries that developers can use to interact with the API and build applications.
- **Swagger/OpenAPI**: An open source standard for describing and documenting APIs.
- **Client applications**: Applications that developers build using your APIs, also known as client apps. When these applications make API calls from the API directory, they are considered client applications of the API directory.
- **Access tokens**: Provided to client applications to grant the authorization necessary to start making API calls.
- **Consumer key and secret**: The credentials associated with a given client application. An access token, which is required to make calls to APIs to which the application has subscribed, is created using the consumer key and secret.
- **cURL**: A command-line program that sends data via different protocols, most commonly HTTP. The cURL syntax is used in a large number of API example calls.
- **eXtensible Markup Language** (**XML**): Designed to store and transport data with tags that say what the data is.
- **JavaScript Object Notation** (**JSON**): In terms of how to transfer and store data, JSON is very similar to XML. JSON is easier to parse and smaller than XML.
- **REpresentational State Transfer** (**REST**): Stateless REST design often uses HTTP. The REST style only allows HTTP verbs to be used for actions on resources (nouns with their own distinct URIs, such as `GET`, `POST`, `PUT`, and `DELETE`). Take the account resource, for instance. To get information about that account, you would use `HTTP GET`, and to update the account, `HTTP POST`. Many people believe that REST-style APIs are simpler to comprehend and use, particularly for mobile application development.
- **Open Authorization** (**OAuth**): An open standard for token-based authentication and permission on the web. OAuth comes in two flavors:
  - **Two-legged**, which just requires the client application to be authenticated
  - **Three-legged**, which requires both the client program and the end user to be authenticated. Right now, two-legged OAuth is supported by all of our APIs. Soon, there will be three-legged support.

- **Simple Object Access Protocol** (**SOAP**): Messages are sent via XML using this web service mechanism, typically as `HTTP POST`. Service actions requested by the calling application are contained in SOAP services. The verb/noun combinations in these service activities (such as `getAccountByID`, `getAccountByName`, and `updateAccount`) are typical. The fact that SOAP offers **Web Service Definition Language** (**WSDL**), which provides a description of the service operations, is one benefit. The majority of enterprise-scale applications provide support for SOAP.
- **SoapUI**: A graphical web service testing tool. It is open source and free. Both SOAP and REST calls are supported by SoapUI.
- **Throttling**: Limiting the number of API calls a given application can perform is achieved via throttling. Most of the APIs in the API directory have a 200-call-per-minute cap.
- **Properties**: Name-value pairs that can be used in the functionality of the API. In various settings, such as sandbox, test, and production, API developers might define a distinct value for the same attribute. The actual value will be used at runtime when the client is consuming the API. The exact API backend URL utilized in each scenario at runtime can be defined using the `endPointUrl` field.
- **Paths**: The resource URLs that API developers use to expose their API. Each path can have a defined `GET/PUT/POST/DELETE HTTP` action and may have predetermined conditions. The parameters can be given as query, path, or header parameters as well as necessary or optional. Additionally, default values for the parameters can be specified by the API developers.
- **Definitions**: The syntax for any properties you want to use when developing the API. In this section, an API developer can specify the behavior of a property, such as whether it is an object or array with particular fields.
- **Tags**: When the API is exposed, tags, which are metadata about the API, will be handy. Tags can be utilized by customers to find the API. The API will appear in the search results if the consumer uses a tag defined in the API. For any given API, we can declare multiple tags.

Knowledge of this terminology will help you understand the context and engage more effectively with customers and stakeholders. It is highly recommended that you explore more tools to develop a deeper understanding of how APIs are used. Knowledge of API concepts would also be useful for you in setting the API strategy, which you will learn more about in the following section.

The short-term and long-term API strategy

Like every product, API products also need a strategy, both short-term and long-term. As you work with market research and user research to identify the market opportunities for your APIs, it's important to define how you plan to shape the product in the short term, along with the milestones that will help you achieve the long-term goals of the product. This would be a combination of identifying the target audience, the marketing strategy, the results you expect in terms of revenue and usage of the product, and also any challenges you foresee in being able to achieve these goals.

As a product manager, being able to communicate the short-term and long-term strategy in a clear way helps align all the teams towards a shared goal, get buy-in from leadership, and get any cross-organizational help you may need. A clear strategy will help the engineering teams to be able to define the infrastructure and architecture needs so that the development of the product can be planned in line with the expected business goals.

Development and release

API product managers are deeply involved with the development and release of the APIs. The development teams working to build the APIs might have the same skills as the development teams using the APIs on the customer front, but oftentimes, the perspective of using APIs and building APIs is considerably different. For this reason, APIs are often launched in a stepwise fashion, where they are developed in collaboration with a handful of customers who can talk directly to the product and engineering teams.

This allows for close partnership, fast learning, and iteration for the API development team. Product managers will coordinate these efforts and evaluate the progress and response of the customer to define the next steps for the product.

API governance and maturity

APIs should follow the certain architecture and design fundamentals to make sure that, over time, the APIs built within an organization and within the same product offering provide a consistent and coherent experience for their users. The guidance that goes into making sure the APIs meet these design standards across the organization is called **API governance**. API product managers would need to ensure that the APIs that are being built meet these standards to make sure the API products are high-quality and meet the standards set by the organization.

In addition to the quality, stability, reliability, and observability of the APIs, product managers must establish and follow a way to establish the maturity of a set of APIs. Maturity is what differentiates an experimental API from APIs that are ready for wide usage. Customers often rely on the maturity label of APIs to make decisions on whether to use an API within their workflow or not. As a product manager, you must ensure that there is a clear definition of maturity for API products established and communicated to customers. Each API has an accurately labeled maturity and, based on the maturity of the APIs, customers have a clear understanding of what level of support (SLA), uptime, reliability, and frequency of change they can expect from the APIs. In *Chapter 3*, you will learn about API governance and maturity in depth.

API experience

The goal of the API experience is to improve API design so that it is easy for developers to use when they are writing software. It can help programmers work more quickly and make it easier for developers to help end users reach their goals.

Developers often discover APIs through a Google search for a specific use case or from blogs, videos, developer forums, and development tools that showcase APIs. The API experience is the term for all the elements of how an API is discovered, evaluated, integrated, and used.

The key components of an API experience include an API reference, an API status page, SDKs, a developer portal, a support path, and third-party development platforms. A developer portal is a website or platform that provides developers with the resources and information they need to interact with and build applications using an API. The API reference is the source of truth for all API-related information and holds details of the API functionality, required and optional parameters, headers, and code samples to get started. The API status page is a page that is updated at all times of day and shows the uptime of the APIs. This is used by customers who are actively using the APIs so that, in the event that there is an incident on the APIs, they can get informed of the status of the APIs. Like any product, it is important that customers have a way to reach support in case they need help in any way.

A great way to learn about APIs is through blog posts that demonstrate the capabilities of the APIs in easy projects that developers can follow along to complete. There are also tools such as Postman that allow you to publish API collections and provide a UI-based experience for customers to discover and get started with APIs without having to write any code. All these aspects make APIs easier to discover and use and are components of the API experience. A great API experience that makes APIs easy and fun to use will ultimately drive the success of the APIs.

In *Chapter 9*, you will learn about the key components of the API experience in depth.

API analytics

APIs, like any other product, have their own set of acquisition, conversion, retention, and usage metrics. Although it might be easy for a developer to try out an API and make a few first calls fairly quickly, oftentimes, revenue is generated from the high-scale usage of APIs. This creates an interesting usage pattern in API analytics that product managers are interested in studying.

As developers discover and evaluate a set of APIs, it might take them anywhere from a few weeks to a few months to complete integration. Product managers will analyze the usage patterns across users to understand how different segments of customers interact with the APIs and how easy or difficult it is for them to integrate. This will help identify opportunities to improve the API experience for specific segments so that they can be more successful using the APIs.

As you evaluate your career in product management, you can review your prior experience against the special skills required for different types of product roles. For example, if you have worked as an analyst or data engineer in your previous roles, your knowledge of data will position you well to become a data product manager. Similarly, if you have a background in sales or marketing, you might be a great fit for a growth product manager role.

Domain knowledge also helps position yourself for success in a new product manager role. For example, if you have worked in finance and are familiar with the financial domain, you can target roles in FinTech or financial organizations. Your domain knowledge of finance along with your understanding of product concepts and methodologies will position you well to break into the product manager role.

In *Chapters 10*, *11*, and *12*, you will learn about establishing API analytics across infrastructure, product experience, and business.

Summary

In this chapter, you learned about the roles and responsibilities of the product manager as well as the skills required to become one. Although the role of a product manager is fairly vast, because of its collaborative nature, it can be quite rewarding. We learned about how, as a product manager, you will bring in various people from across the organization and from various areas of expertise to make a product that serves customer needs. You will get to work with experts in user research, design, development, sales, and marketing to unlock market opportunities and deliver innovative products.

We also dove into the unique responsibilities of the API product manager and how these complement the foundational role of a product manager. In the following chapter, we will dive deeper into concepts such as API governance and maturity.

3

API Life Cycle and Maturity

At this point, you have now learned about and are familiar with the various types of products, including B2B products, B2C products, platform products, and API products. You have also learned about the role of a product manager and how different types of product managers have a wide range of commonly-found skills alongside a set of particular skills unique to their specialization. In this chapter, we will focus on certain aspects of API products that are critical for API product managers to understand so that they can analyze the API products available on the market and help shape the API products they lead.

In this chapter, you will look at some case studies of the most prominent API companies to understand how they establish and communicate their API life cycles and maturity to their customers, covering the following key aspects of APIs as a product:

- API product life cycle
- API governance
- API maturity

The API product life cycle

In *Chapter 2*, you learned about the product life cycle, where you take product ideas from discovery, inception, growth, maturity, and decline. When you apply that concept to APIs-as-products, you get an API product life cycle. APIs are developed in an iterative way, with features focused on the first set of customers who will use them. The best way to learn about customers is to get them to engage, and most of the time, API development teams will do just that.

At first, like any other product, there might just be an idea for building an API. This is a good time to put together an API proposal to evaluate the use cases that the prospective API could serve in order to establish the features of the first iteration of the API.

An API goes through six phases in its life cycle, from the time of inception to the state of being a generally available product ready for all customers to use and trust. Once an API has been published, it must be maintained so that customers can use it reliably. Eventually, it's possible that APIs may also be deprecated or retired to be replaced by a newer version, and this version would go through the six phases of being published.

The following infographic outlines the API product life cycle:

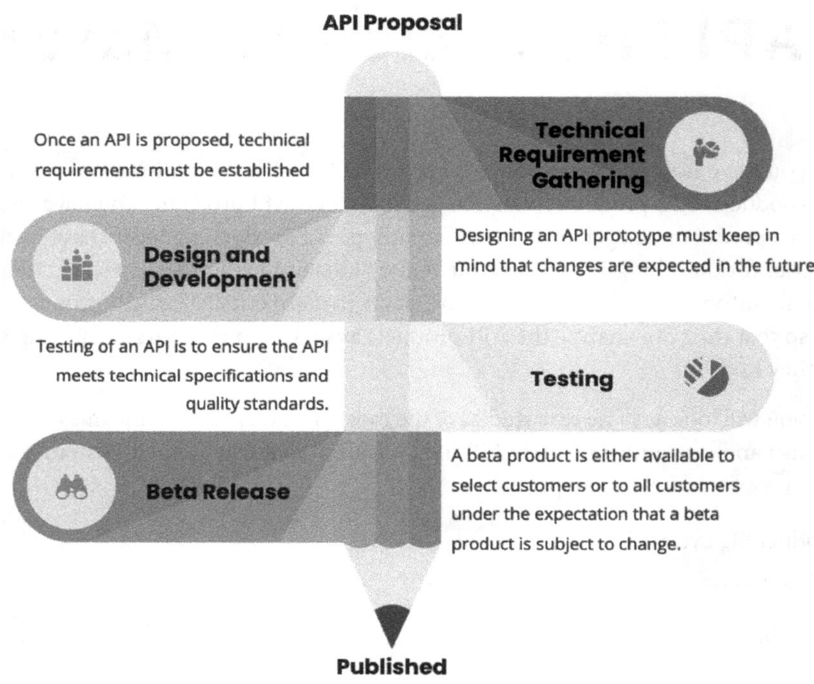
API Proposal
Technical Requirement Gathering
Once an API is proposed, technical requirements must be established
Design and Development
Designing an API prototype must keep in mind that changes are expected in the future
Testing
Testing of an API is to ensure the API meets technical specifications and quality standards.
Beta Release
A beta product is either available to select customers or to all customers under the expectation that a beta product is subject to change.
Published

Figure 3.1 – API product life cycle

The infographic shows how an API concept moves from the early stages of gathering requirements, designing, and developing, to testing, to being released to a small group of users as a beta version, to being published and made available to a wider audience. Let's look at each of these stages in more detail:

1. **Proposal**

   Before development starts, an API specification must be written by the product manager in collaboration with their engineering counterparts. This document will serve as the foundation of the product and establish why this API is needed, who the ideal user is, what user problem this API solves, and how this API meets the business goals.

Product managers often write a **PRFAQ** (**Press Release Frequently Asked Questions**) document or a one-pager to propose a new product. This addresses why this product should be built, who it is targeted at, and identifies the business opportunity. This document will then be presented to the product and engineering leadership to make the case for investing development resources in this product.

This is where the strategy for the API needs to be defined in terms of what business goal the API serves and how it serves this goal. For example, APIs may or may not be the primary product offering of a business, but an API offering might enable integrations that in turn enable the onboarding of more customers. Like any other product, it is important to evaluate what metrics a product is targeting and how.

Having a clear strategy will also help define the milestones ahead in the product development journey and establish the short- and long-term goals for the product.

Above all, determine for what purpose a proposed API needs to exist. This will help you figure out what problems it aims to solve and how big of an effect it could have. This also helps the rest of the team understand the big-picture strategy for the API, and this can help the technical architects design solutions that can scale to serve the use cases that are the focus of this effort.

2. **Technical specifications**

   Based on the product specification, engineering teams then create technical specifications for the API they plan to develop. This could be based on existing infrastructure for APIs in the organization, or if this is the organization's first API, it could involve the creation of new infrastructure. If the organization is at the stage of building its first APIs, then it needs to review its organizational readiness and create a plan accordingly.

3. **Design and development**

   The design and development phase of the API life cycle is where the developers start to prototype the API. Architects might be brought in by the engineering teams during the design phase to make sure that the API design meets the standards set by the architecture team.

4. **Testing**

   Once a functioning prototype is ready, the testing and development cycle begins, where the API goes through rigorous testing scenarios, and any bugs. The gaps discovered in this process must be put back into the engineering backlog for review.

5. **Beta release**

   APIs are not directly launched on a global scale, but usually go through a gradual rollout process. A beta release signifies that the product is still in development and subject to change. Beta releases can be further broken down into internal beta and private beta releases. An internal beta release is when internal teams are invited to test out and integrate the API. A private beta is usually rolled out to external customers on an invite-only basis, focusing on those who are more engaged with the development team and can actively provide feedback. This phase also allows product teams to validate the API use case and facilitate learnings that shape the API.

6. **Published**

   When an API is *published*, it is available for public use. At this point, the API is accompanied by appropriate API references, status pages, and changelogs. Companies that publish public APIs also give a maturity status to their API to communicate the SLAs explicitly and manage customer expectations. A mature product is expected to be reliable, robust, and scalable and the maturity level stated by the company helps customers evaluate the APIs for their use.

As your API goes through its product life cycle, you might have many development iterations that go from prototypes to mature products. Each iteration of the product contains several steps that product managers may or may not be directly involved with. In the following section, you will learn about the API development workflow that engineering teams use to produce APIs.

API development workflow

As a product manager, you will be focused on the API product life cycle. However, there are a number of steps within the technical requirement gathering, design, development, and test phases of the API product life cycle that each API goes through at an engineering level within an organization before APIs are ready to be distributed to users. These steps form the API development workflow.

In an enterprise setting, when a product manager presents a proposal for an API, the engineering lead augments the business requirements with technical and security requirements for the API's development. The engineering team also identifies the right workspace for the API and sets up source control before the development begins.

The process that engineering teams use to make APIs is shown in the following diagram. The API development workflow includes a number of tools and technical stakeholders across the stages of definition, design, development, testing, security, deployment, and observability. Once the API is ready for users, it is distributed.

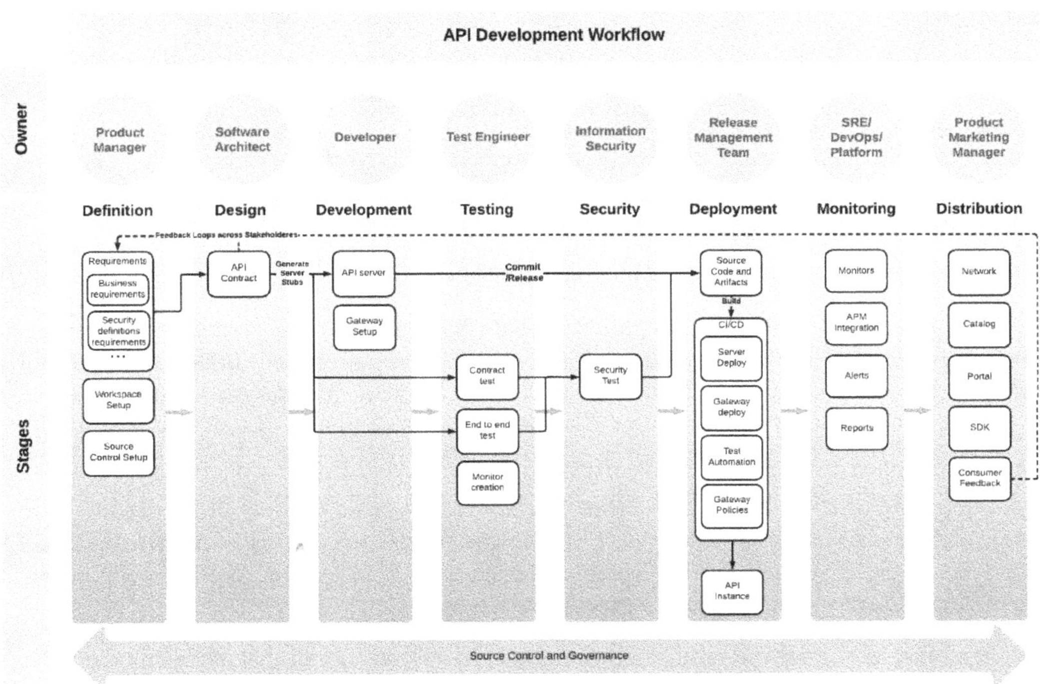

Figure 3.2 – API development workflow

Once your API proposal has been reviewed and approved, you will work with the software architects and developers on your team to get the API into the design and development phase. Product managers aren't expected to be too involved in the development process, but it's important to know how engineering teams work so that you can work with them effectively. Familiarity with the API development process, the tools involved, and the stakeholders will help you navigate cross-team communication and manage stakeholder expectations.

Let's now discuss in more detail the eight stages of the API development workflow shown in the preceding diagram.

Definition

The definition stage involves the product manager working with their engineering leads and software architects to put together the technical, business, and security requirements for the API.

At this stage of the API life cycle, operations are specified. This lays the groundwork for designing and implementing an API efficiently and provides a central location for all API-related information.

Make sure to establish who in the team will move the API through all stages of its life cycle, defining who is responsible for each phase of producing an API.

Setting up a separate workspace for each API makes sure that there is a single place to find everything that is being worked on during the development life cycle. With a dedicated GitHub repository for a given API, all the code, pipelines, problems, and other supporting features can be managed from one place.

Design

Software architects review proposals for new API designs and generate the API contract. These API contracts can be used to generate preliminary documentation and mocks using tools such as Postman. This provides an early view of what the API will look like.

The design phase of the life cycle establishes a formal process and approach to developing an API, ensuring that APIs are designed using established industry and organizational patterns and known practices for defining the API surface area and behaviors.

The OpenAPI Specification provides a uniform vocabulary for specifying request and response APIs and webhooks, ensuring API providers and consumers are on the same page. It helps stabilize the API life cycle by providing a contract for documentation, mocks, testing, and more.

A mock server helps duplicate an API's production capabilities before any code has been written, making APIs more concrete and real earlier in the life cycle. This helps teams to quickly develop APIs that satisfy everyone's expectations, offering useable elements for design, documentation, and testing.

Development

The development stage refers to the actual setup of the API server and gateway. PHP, Python, Ruby, .NET, C#, Java, and Perl are all API languages. Coding and testing may require revisions, but developers shouldn't deploy an unstable API. API development requires planning, coding, testing, and the discipline to limit customer-facing API modifications.

API development can be done by a single developer or a professional team. A team approach lets multiple developers establish and maintain the organization's API portfolio. Multiple developers make it easier to write public API documentation, test cases, and marketing materials.

Multiple developers must access and manage their APIs together. API code, documentation, and test files should be accessible to the entire team. Implementing sophisticated search functions and version control enables authorized developers to find APIs by project and version.

When developing an API, there are several key considerations to keep in mind:

- **Tracking API use**: It is important to track how the API is being used in order to understand how it is being adopted and identify any potential issues. This can include tracking the number of requests, the types of requests, and the response times.
- **Performance**: The API should be designed to handle a large number of requests and return responses quickly. This may involve optimizing the database queries and caching responses.
- **Failures**: The API should be designed to handle failures gracefully and return appropriate error messages. This includes handling failed requests, failed authentication, and failed authorization.
- **OAuth 2.0 Authorization**: OAuth 2.0 is an authorization framework that allows users to grant third-party applications access to their resources without sharing their passwords. It is widely used in API development to ensure secure access to protected resources.
- **API Key validation**: API keys are used to authenticate and authorize API requests. It is important to validate API keys on the server-side to ensure that only authorized clients can access the API.
- **Throttling and rate restrictions**: To ensure API performance and accessibility, it is important to set throttling and rate restrictions. This can include limiting the number of requests per second or per minute, or limiting the number of requests per user or per IP address. Development goes hand in hand with testing which you will learn about in the next section.

Testing

An API test is generally performed by making requests to one or more API endpoints and comparing the response with the expected results to make sure the API meets the required functionality, security, performance, and reliability expectations. There are a variety of API tests ranging from general to specific analysis of the software, including the following:

- **Performance testing**: is the process of evaluating how well an API performs in terms of responsiveness and stability under a certain workload. This can help identify bottlenecks and optimize performance.
- **Load testing**: is a type of performance testing that involves subjecting an API to a heavy load, such as a high number of concurrent users, in order to determine how well it can handle large amounts of traffic.
- **Functional testing**: is the process of testing that an API performs as expected and returns the correct output for a given input. This helps ensure that the API is working correctly and is able to handle different input scenarios.

- **Reliability testing**: is the process of evaluating how well an API can handle unexpected situations and recover from errors. This helps ensure that the API is robust and can continue to function even in the event of unexpected conditions.
- **Security testing**: is the process of evaluating the security of an API to identify vulnerabilities that could be exploited by attackers. This includes testing for things like SQL injection, cross-site scripting, and other common security threats.

API testing allows developers to start testing early in the development cycle before the user interface is ready, and kill at least half the existing bugs before they become more serious problems, which saves money in the long run. API testing is critical because it guarantees the connections between platforms are reliable, safe, and scalable.

Security

During the security phase of the API life cycle, teams are responsible for ensuring that all APIs follow the same rules when it comes to managing user identities. As an added precaution, rigorous security testing is implemented to ensure that all APIs are scanned and secured in the same manner, regardless of which team built the API or whether it will be used privately, with partners, or publicly in applications.

Authentication and authorization are two separate but related concepts when it comes to APIs.

Authentication refers to the process of verifying the identity of a user, system, or device that is attempting to access the API. This is typically done by requiring a user to provide a set of credentials, such as a username and password, that can be verified against a database of known users.

Authorization refers to the process of determining whether an authenticated user, system, or device has the appropriate permissions to perform a specific action or access a specific resource within the API. This is often done by comparing the authenticated user's roles and permissions against a set of predefined rules, such as access control lists or role-based access control.

The **Open Web Application Security Project** (**OWASP**) is a nonprofit organization that aims to improve the security of web applications by providing resources and tools for developers, security professionals, and educators. OWASP has a specific project for API security, which provides a wealth of resources for API security including best practices, testing guides, and tools. These resources can help developers understand the specific security risks associated with APIs and how to mitigate them.

Deployment

To help teams effectively deliver future iterations of an API in a consistent and repeatable manner, we must focus on establishing a well-defined procedure to deploy APIs into development, staging, and production environments. This makes testing, security, and governance of APIs an integral component of the deployment process from the get-go.

To successfully publish APIs in development, staging, and production environments, you need a **continuous integration/continuous delivery** (**CI/CD**) pipeline, just like you would for any other part of the software development life cycle.

API gateways use commercial or open source solutions to deploy APIs into development, staging, production, or any other environment, allowing for centralized or federated management of APIs at scale throughout an organization. This makes it possible to standardize the different parts of API administration and set up APIs so that they follow the same rules.

Monitoring

Monitoring is an important aspect of the API development lifecycle because it helps ensure that the API is functioning as expected and that any issues are quickly identified and resolved. Monitoring is typically done after the API has been deployed and is in production. Monitoring can help in various ways:

- Performance monitoring: By monitoring the performance of the API, developers can ensure that it is meeting the expected response time and throughput. This can help identify bottlenecks and optimize performance.
- Error monitoring: By monitoring for errors and exceptions, developers can quickly identify and address any issues that may be affecting the functionality of the API.
- Security monitoring: By monitoring for security-related events, such as attempted breaches or unusual activity, developers can quickly identify and respond to potential security threats.
- Usage monitoring: By monitoring usage of the API, developers can understand how the API is being used and make adjustments to improve the user experience and optimize the API for its intended use case.
- Compliance monitoring: By monitoring the API for compliance with relevant regulations and standards, such as data privacy laws and industry standards, developers can ensure that the API is meeting all legal and regulatory requirements.

Monitoring helps ensure that the API is operating as intended, and that any issues are identified and addressed quickly, which helps maintain the reliability and security of the API, and help to improve the user experience.

At this point, tests for contract compliance, performance, security, and other metrics are scheduled across different cloud regions. Notifications help teams and users stay abreast of the status of the APIs on which they rely by tracking activity, tracking changes, and monitoring the API's current state being up or down.

For all API contracts, you can set up a monitoring system to run tests at set times and from the locations that are most important to your business. That way, you can rest assured that your API is always meeting its contractual obligations to its users.

Distribution

As the development teams complete the development, testing, and deployment of the APIs, the sales and marketing teams begin to prepare for the distribution of these APIs to customers. Developer relations teams are also instrumental in getting APIs to customers. Careful distribution of APIs is a process that needs collaboration from all the teams involved to make sure the right customers are reached and that there are appropriate feedback loops set up.

These steps ensure that the APIs being built across the organization meet the design and development standards set by the organization. The standards that must be followed at each step of this process are set by API governance, which you will learn more about later in this chapter. This is done to make sure that the APIs being built will last, be scalable, and be reliable. As you can see in *Figure 3.2*, there are several stakeholders involved at various steps of the API product life cycle and API development workflow, and that is why stakeholder management and alignment are key aspects of API product management, as you will learn in the following section.

Stakeholder alignment

Internal APIs have internal customers, meaning other teams inside the organization. Since a given API is being built for mutual benefit, it is easy for the API teams to engage with each other, communicate their needs, and deliver value.

For external APIs, the customers are external to the organization and may not be as easily engaged in the development process of the APIs. If the goal is to drive adoption across a wider range of audiences, it is also important that the design not be over-fitted to the needs of a handful of customers.

In addition to the customer who will ultimately use the APIs, there are a number of internal stakeholders who have a say in how the APIs are built. It is important for a product manager to recognize where in the life cycle a product is so that the right stakeholder expectations can be set in terms of metrics and business decisions. These could be decisions such as whether a product is ready for sales teams to actively sell it to customers, or whether a product is still in the development phase and so it might be too soon to sell to customers. If a product is in beta, not every customer will feel comfortable onboarding to the product because beta products are usually subject to change, and customers may not have the necessary developer bandwidth if they need development changes to be made.

Stakeholders for API products include the sales, account management, and support teams, all of whom may be communicating with customers on a regular basis and be deeply aware of their needs. In the early phase of API development, sales teams can help connect with the right customers to get feedback and evaluate customer pain points. Other stakeholders include the legal and information security teams who are interested in making sure the APIs being proposed are secure and compliant. Architects within the organization are interested in making sure the proposed APIs are designed in a way that they fit together with other APIs offered by the company and meet the design standards laid out from an architectural perspective. The product marketing team is usually involved at the time of launch to make sure the right messaging and marketing strategy is put in place for the product to be successful.

The stakeholders who are interested in knowing about where an API is in its lifecycle could include sales team members or account managers who are working with a customer who has requested certain functionality, the support team who will be providing customer support for your API when it is launched, the risk and compliance teams who need to evaluate that the products being developed meet regulation, the partner teams who might be looking to use your API to enable their product, etc. In startups, the CEO and other C-level executives might also be involved in the development process as every product is tied closely with the company's growth and it is important that these stakeholders are aligned in their expectations of your API

Retiring an API

The development life cycle of an API describes the continuous evolution of the product, but at some point, just like any other product, APIs become outdated and need to be retired. Over time, continuous customer feedback, support costs, scalability challenges, and other concerns might lead to the redesign of an API. There are also security concerns that can lead to the decision to close down a product offering.

Whenever a product retirement decision is made, product managers have to put in place a clear strategy to reduce the impact on the customers as much as possible. It is not possible to simply discontinue an API because customers might have built their applications using these APIs and some customers may be unable to align their roadmaps with the timeline of deprecation. Before an API can be retired, it is first placed in a deprecation status.

Since your customers are likely actively using your APIs in their applications and have a business need for them, you must implement the following best practices for retiring APIs:

- Track the current usage of the APIs in terms of the number and percentage of customers using the API that you plan to retire. This list of customers can be used to make sure you reach them to inform them of upcoming changes.
- Identify a replacement solution for the APIs being retired. If there is a functionality that is being discontinued, you should ensure there is an alternative path that you can recommend to your users as a solution to switch to.
- In a B2B setting, you can partner with sales, account management, and customer success teams to get in touch with customers early to understand their dependency on the API that is being planned for retirement. This allows you to set a timeline for API retirement with the least impact on the customers. Most commonly, at least two quarters is recommended to be communicated to customers that the APIs they use are on the path to being retired, so that they can plan the necessary migration on their end. You must also consider that most enterprises make annual IT plans and allocate budgets at particular points in the year, and B2B API retirement plans need to accommodate this.

- Continuous reminders are key in planning an API deprecation. You must work with customer success teams closely to ensure customers are not taken off guard when the APIs are deprecated. Repeated communication ensures customers are informed and can plan dependencies effectively without losing sight of your notification.
- Having updated customer-facing changelogs that document the upcoming changes for customers is crucial. Depending on how much emphasis is needed, you can complement your changelog posts with posts on social media and blog posts. If the impact on customers from the API being retired is not major, you may not have to make too many announcements and a changelog might be sufficient, but if you suspect that the impact is going to be substantial, you should ensure extensive customer communication.
- Placing any product on the path to deprecation requires both planning and execution in a way that the customers are not inconvenienced and the impact on business is minimized. To this end, you can partner with all the stakeholder teams to use their areas of expertise to drive a smooth transition for the end customer.

Now that you understand how various stakeholders are involved in the API development workflow, you will learn about standards, best practices, and SLAs that allow these stakeholders to operate this workflow as part of an effective API governance model.

API governance

There is a growing number of companies building APIs. As the number of APIs grows within an organization, consistency becomes a key challenge because external customers approaching these APIs from outside the organization expect a consistent experience and expect all APIs to be compatible and have tight coupling among all components. Inconsistency between different API offerings can lead to customers being confused and increase the time it takes them to integrate the APIs.

API governance is a term coined to describe an enterprise-level set of standards for the design and operation of APIs. APIs can be built in a variety of ways and because of the iterative API development process, it is important that certain design patterns be established early in the life cycle so that APIs can be built across the organization in a way that fits together predictably and reliably.

For APIs to be trusted in terms of safety, availability, scalability, and dependability, API governance must set standards and principles for each of these qualities. To design strong APIs and make sure developers have a great experience, it's important to have a policy-driven strategy that helps enforce standards and checkpoints throughout the API's life. This strategy will explain the rules and procedures that must be followed during API discovery, interface documentation, development, testing, deployment, operation, and maintenance.

API governance through the API life cycle

Once you understand the API life cycle, it is easy to imagine that with numerous teams and stakeholders involved, there is a need to enforce the API governance standards at each stage of the API development life cycle. This is accomplished as follows:

- When an API is proposed, the process of API governance begins, and it is the responsibility of the development and operations teams to verify that the APIs they identify are consistent with the overall company plan. An API proposal, like any product proposal, must make a good case for investing in the development of the product and be able to showcase the value it will drive for the business. This could involve stakeholders from across the organization including sales, legal, and support to ensure there is buy-in from these teams and they align their roadmaps accordingly.
- In the design and development phase, API governance plays a key role in ensuring that the API software quality is maintained and that the user experience is in line with other APIs published by the organization to enable a smooth onboarding for both existing and new customers. To ensure that the API's quality lives up to the standards required by the company's customers, a solid API versioning strategy and emphasis on development standards are essential.
- To ensure that the necessary expectations of security, reliability, availability, and scalability are met, an extensive API testing strategy is established so that the quality of the product can be measured and **service-level agreements** (**SLAs**) can be established and enforced.
- API runtime quality drives customer trust, so API governance must set standards for API monitoring, deployment, and dynamic provisioning to guarantee API runtime quality and SLAs that can be communicated to customers clearly and reliably.
- Once SLAs are established for API governance, these SLAs must be monitored to ensure that they are met by the API provider and the consumer.

When APIs are developed with appropriate governance rules and practices, effective measures and controls can be established to understand how ready your APIs are for customer use. In the following section, you will learn about API maturity and how to measure an API's readiness for customer use and enable a controlled release process.

API maturity

API governance is utilized to drive consistency and trust through the development of APIs, but most of the work that teams do on API governance prior to APIs being published is completely hidden from the end customer. Like all great designs, the customers only experience a polished product and don't notice the thoughtful details in the background that make it all possible.

Although the internal stages of the API life cycle are hidden from the customer, what is visible to the customer is API maturity. API maturity is a label that companies give to their APIs to set expectations with the customer regarding supportability, versioning, and backward compatibility, among others.

The following diagram illustrates the different levels of maturity that an API goes through as it gets developed and is released to customers:

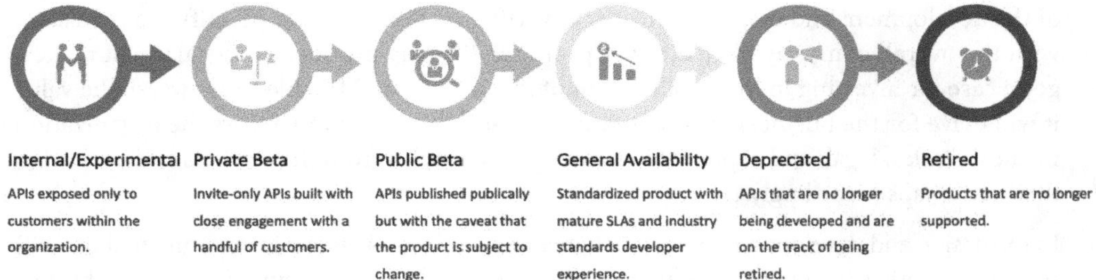

Internal/Experimental
APIs exposed only to customers within the organization.
Private Beta
Invite-only APIs built with close engagement with a handful of customers.
Public Beta
APIs published publically but with the caveat that the product is subject to change.
General Availability
Standardized product with mature SLAs and industry standards developer experience.
Deprecated
APIs that are no longer being developed and are on the track of being retired.
Retired
Products that are no longer supported.

Figure 3.3 – API maturity levels

Companies building commercial APIs, such as eBay, Twilio, Okta, Stripe, and others, drive customer trust by communicating clear SLAs, release cycles, and versioning in their API documentation. Making APIs robust, reliable, and scalable requires emphasis on the infrastructure side of API development. This aspect, however, is completely invisible to the end customer, who simply has to assume that the infrastructure is kept to a high standard. Because the customer may not have a way of evaluating the robustness, reliability, and scalability of a given API, the API maturity status becomes the proxy for the customer to interpret these API characteristics so that they can make decisions on using the API for their development.

API maturity communicates the following elements to customers:

- The performance and scalability of the API
- SLA expectations for each level of maturity
- Degree of backward compatibility for upcoming features and expectations of changes in API functionality
- Documentation standards
- Ownership and liability in the case of incidents and when SLAs are not met

API maturity is closely aligned with the release process of an API. When you roll out new features in a way that they are first tested internally to make sure they are working as expected, you can feel confident that features are working as expected. New features are then carefully rolled out to wider audiences in a beta version to set expectations and provide the opportunity to make improvements based on the feedback that is received at this stage. Once the products are performing at close to general availability capabilities, only then are they transitioned to general availability. This method ensures that products are released in a step-by-step way and that only high-quality products are labeled as being available to the public.

The controlled release of new products and features is a methodology that allows quality control and measured improvements to be made. When you keep the quality of products that are available to everyone high, you can set SLAs that build customer trust. In the next section, you'll learn how the most well-known API-producing companies determine maturity levels and communicate them to their customers.

Case studies

Even though the API life cycle is well known and established, different companies define API maturity statuses in different ways. In the following section, we will take a look at some case studies of API maturity and see how it is defined and communicated by some of the most prominent API-first companies.

Twilio

We looked at Twilio in *Chapter 1*, where we learned about APIs as a product. Twilio is an API-first company and has a sophisticated way of presenting the maturity model and SLAs for their APIs very clearly on their website. Any type of SLA breach has a revenue impact in the form of service credit. In the following screenshot, you can see how Twilio communicates its SLAs to its customers to set clear expectations.

Applicable Services	Monthly Uptime Percentage Threshold	Service Credit
"Services" as defined in the Agreement	99.95%	10% credit equivalent
During calendar months in which Customer has purchased the Twilio Administration Edition or Twilio Enterprise Edition	99.99% for Twilio Services	10% credit equivalent
During calendar months in which Customer has purchased the Twilio SendGrid Enterprise Solution	99.99% for SendGrid Services	10% credit equivalent

Figure 3.4 – Twilio's Publicly Communicated SLAs across the different APIs they offer

Twilio has three maturity levels of SLA:

- **Private Beta**
- **Beta**
- **General Availability** (GA)

For APIs with GA status, the committed uptime percentage threshold is 99.95% based on level, and in the event that Twilio doesn't meet this SLA, the company commits to providing customers with a service credit of 10%. For transparency, API uptime metrics are published on `status.twilio.com` where customers can go to check when the Twilio APIs are down. This page is kept updated at all times. Twilio also sends out API status and incident notifications through a Twitter handle. This lets customers know whether an incident occurs that could affect their integration so they can fix it quickly.

For customers using Twilio APIs and subscribing to paid support, they can expect support as quickly as 1 hour for business-critical issues, while developers not subscribed to paid support could wait from 24 hours to 7 days. This degree of detail helps customers understand what they can expect from Twilio in terms of the reliability of the product and support in the event of issues. Twilio also ensures that if and when customers encounter an incident, they learn about it sooner rather than later and communicate the impact to their customers accordingly by broadcasting API status updates on various social media channels.

Block

You might have heard of the fintech company **Square**, now called **Block**. The Block card reader first appeared in 2010 and took the small-business-payment space by storm. Since then, they have expanded into all kinds of payments, order processing, and small business management offerings. The Block team has done a great job of putting together a clear versioning and release notes process in place. Block's API life cycle is documented on a dedicated page at `https://developer.squareup.com/docs/build-basics/API-life` cycle. The Block APIs are labeled as follows:

- Beta
- General Availability
- Deprecated
- Retired

The following screenshot shows how Block presents its schedule for API retirement and deprecation in its developer documentation:

Endpoint name	Replacement	Deprecated	Retirement
RenewToken	OAuth.ObtainToken Migration guide	2021-05-13	2022-05-13
ListEmployees	ListTeamMembers Migration Guide	2020-08-26	2021-08-26
RetrieveEmployee	RetrieveEmployee Migration Guide	2020-08-26	2021-08-26
ListEmployeeWages	ListTeamMemberWages Migration Guide	2020-08-26	2021-08-26
GetEmployeeWage	GetTeamMemberWage Migration Guide	2020-08-26	2021-08-26
ListLoyaltyPrograms	RetrieveLoyaltyProgram Migration Guide	2021-05-13	2022-05-13
CreateCustomerCard	CreateCard or CreateGiftCard Migration Guide	2021-06-16	2022-06-16
DeleteCustomerCard	DisableCard or UnlinkCustomerFromGiftCard Migration Guide	2021-06-16	2022-06-16

Figure 3.5 – Block's publicly communicated deprecation and retirement status and timelines for APIs

The documentation for API versioning is always kept up to date, and release notes are carefully written to tag specific APIs. Beta APIs are expected to have minor changes between beta and GA, but not major changes. Block also specifies that when APIs are deprecated as the first stage toward retirement, they stay in the deprecated state for at least 12 months before reaching permanent retirement.

This level of clarity sets customer expectations really well in terms of the usability of these APIs. Transparency in policy and communication can drive customer trust and also help customers self-serve, also saving customers time.

Block presents a current status page for each API in their documentation, the details of each endpoint, and the schedule for deprecation and retirement to enable customers to plan ahead in case any APIs they depend on are on the path to deprecation or retirement.

Okta

Okta is a cloud identity platform that allows customers to integrate with their platform and manage **single-sign-on (SSO)** and multi-factor authentication.

If you look at Okta's API documentation, you will notice that the company does a great job of communicating in clear terms how to interpret the maturity status of its APIs. Okta works to set its customers' expectations so they can make informed decisions about which product is at the right level of maturity to be used for their production workflows.

The Okta maturity model is presented in a very concise table as follows:

Quick Reference Table

Description	Beta	EA	GA	Deprecated
Contact with Product Team	✓	✗	✗	NA
API Changes	Subject to change	Backwards compatible	Backwards compatible	N/A
Okta Support	✗	✓	✓	✓
Service-level agreements	✗	✓	✓	✓
Announced in Release Notes	✗	✓	✓	✓
In preview orgs	By invitation or self-service, depending on the feature	By request or self-service, depending on the feature	✓	✓
In production orgs	✗	By request or self-service, depending on the feature	✓	✓
Documentation	Limited	✓	✓	NA

Figure 3.6 – Okta's publicly communicated API maturity levels and respective SLAs

The Okta API maturity life cycle consists of three steps:

- **Beta**

  Beta APIs and features are only available to customers by invitation. Okta utilizes a newsletter to inform customers and communities about betas.

  While a beta API is active, users can anticipate frequent communication with Okta, such as conference calls to discuss use cases, deployment instructions, and feedback for the API development team.

  Some beta products may be self-service and lack Okta support or contact. Any team communication is for feedback. Beta features can be enabled from the **Account** > **Features** tab in the Admin Console.

- **Early Access (EA)**

  EA features are new or upgraded functionalities that users can opt in to and utilize in production and non-production contexts.

- **General Availability** (GA)

  Any new or improved functionality that is made available to all users by default is said to be in GA.

This level of detail makes sure that customers can use Okta to design their apps in a reliable and predictable way.

Shopify

Shopify Inc. is a Canadian company with a full SaaS e-commerce platform offering. Its proprietary e-commerce platform for online stores and retail point-of-sale systems is accompanied with detailed documentation, release schedules, and communication channels for developers. Shopify presents its API release notes in its developer documentation at `https://shopify.dev/api/release-notes`. This documentation is designed to make it easy for developers to get important information in a clear and straightforward way as follows:

- The Shopify developer documentation provides detailed information about any breaking changes as part of new releases, along with which APIs will be impacted
- A clear release schedule with timelines is published to communicate upcoming APIs to customers
- The documentation identifies when APIs are ready for use in development versus production to avoid customers running into issues and helping them plan their development cycles effectively
- The documentation offers links to every possible relevant document page so that users can navigate effectively for more information

Shopify has a variety of REST and GraphQL APIs. As the company expands its API offering, it keeps to a consistent, easy-to-follow organization of information, always starting with *get started* guides to help customers get started with their APIs as easily as possible.

The Microsoft Graph API

Microsoft Graph categorizes API maturity as **Beta**, **GA**, or **Deprecated**, and sets customer expectations for when an API is ready for widespread use. There are currently no deprecated versions of MS Graph.

Beta version

APIs are first launched in beta version and are accessible via `https://graph.microsoft.com/beta endpoint`. Beta API documentation is provided in a dedicated location here: `https://docs.microsoft.com/en-us/graph/api/overview?view=graph-rest-beta&preserve-view=true`. The documentation makes clear to customers to expect breaking changes and the deprecation of beta APIs from time to time. Beta APIs are not recommended for use in production.

The documentation clearly states that it is not guaranteed that beta APIs will be promoted to current version status. When the Microsoft Graph API team considers a feature as ready, it is added to the current version and made available to everyone.

If new features are added to the current version and start to cause problems, the version number goes up by one. This makes sure that customers who are still using an older version don't have to deal with changes that break their integration and that they can upgrade their integration to the latest version according to their development cycle.

The **Microsoft 365 Developer Platform Ideas forum** (`https://techcommunity.microsoft.com/t5/microsoft-365-developer-platform/idb-p/Microsoft365DeveloperPlatform/label-name/Microsoft%20Graph`) allows the developer community to submit feature requests, such as requests to promote beta APIs to the current version.

Current version

Microsoft Graph has reached version 1.0. You can find all of the features that are ready for production use in the Microsoft Graph API v1.0 at `https://graph.microsoft.com/v1.0`.

Preview status

The **(preview)** label shows that an API or feature in Microsoft Graph has special features only available on the beta endpoint. Most APIs and functionality in v1.0 operate the same as they did in beta. There are two types of APIs and features that fall into the **preview** category:

- Those only in beta testing
- Rather than v1.0, it is available in beta

All APIs in the beta endpoint, even those labeled **preview** in the documentation, can change quickly and without warning. Do not use the beta endpoint in production applications.

For example, the Microsoft 365 Defender portal has an attack-simulation training module for admins. The documentation for the attack-simulation training REST API will start with **(preview)** to show that it is currently only available on the beta endpoint of Microsoft Graph. The Microsoft Graph REST API and its documentation are marked as **(preview)**, even though the service is available to the public.

ebay

eBay is one of the world's largest e-commerce platforms and has countless integrations to support a vast and complex ecosystem. In the eBay API documentation, you will find that eBay APIs have three release stages in addition to the **Limited Release API** (invite only), **Experimental APIs**, and **Deprecated APIs**. eBay APIs have one of three maturity statuses: **Alpha**, **Beta**, or **GA**, and each of these is defined by the following properties to help customers understand what to expect:

- Purpose and benefit
- Permissions
- Level of support provided
- API design stability
- Quality
- Intended use

In the following table, you can see how eBay communicates its API maturity schema. It also provides info on whether the API can be used for onboarding customers or load testing. This helps customers know what to expect.

Characteristic	Launch Stage		
	Alpha	Beta	General Availability (GA)
Purpose and benefit	Purpose is to test and develop API features and functionality requirements. Opportunity to preview and influence future eBay API features.	Purpose is to release API capabilities for feedback. Opportunity to get early access to new APIs before GA.	Purpose is to release API features for production use.
Who can access	Available to select eBay Developers Program Members with NDA.	Available to all eBay Developers Program Members, may be subject to eligibility criteria.	Available to all eBay Developers Program Members, may be subject to eligibility criteria.
What support is provided	Support is provided by account management support only, and may not have formal documentation	Standard technical support channel available, with draft documentation.	Standard technical support, with full documentation available.
API design stability	Changes to API likely from feedback in Alpha testing, and Beta may not be backward compatible with Alpha.	Changes to API possible during Beta period and GA may not be backward compatible with Beta.	Stable Interface Design
Quality	May have significant design and availability issues.	May have design and availability issues.	Production
Intended use:	Test environments or limited-use testing.	Limited production use but not for business critical use.	Production
Usable for customer onboarding?	No	Yes	Yes
Usable for load testing?	No	No	Yes

Figure 3.7 – eBay's publicly communicated API maturity levels to set customer expectations

This level of clarity should allow customers to make decisions based on the APIs' maturity status, allowing them to choose the level of maturity that is appropriate for their use case.

Now that you have a good understanding of how some of the most prominent companies present their API maturity, you can design maturity models that suit the nature of your own APIs and business use cases that best suit your customers' needs. Communicating expectations to customers in a clear and transparent manner enables them to make decisions quickly and effectively. The goal is to make the required information as easy as possible to understand and to remove any ambiguity.

Summary

Now that we have had a thorough examination of the API development life cycle, API governance, and the various API maturity models established by key API-first companies, you should have a good idea of how you can define these for any APIs that you plan to produce for external consumption. An understanding of industry standards will help you take into account how your API customers approach this information and which key pieces of information are required to help them make the decision to use an API product.

In the following chapter, we will examine how to develop and manage API products that involve working with various stakeholders and handling products through the various release and maturity states that we learned about in this chapter.

4

Building and Managing API Products

In previous chapters, we learned about viewing **Application Programming Interfaces** (**APIs**) as products and how maturity and governance are established across the API life cycle. In this chapter, we will start to dig deeper into the process of building and managing API products and how to phase this process so that we can do so methodically and systematically.

As you advance an API product from an experimental product to a more mature product, you must continue to develop the product strategy to keep up with evolving customer expectations and stakeholder needs. In this chapter, we will look at the various aspects of building and managing API products in line with these evolving needs as the API goes through its maturity life cycle.

We will cover the following aspects in this chapter:

- APIs in the digital value chain
- Leading with the API strategy
- The API development team
- Managing an API as a product
- The API proposal
- Designing APIs
- Starting with an MVP
- Building long-term and short-term roadmaps

APIs in the digital value chain

The API team should be viewed as a refinery that processes raw materials from backend systems into API products, which provides developers with clean, well-organized ways to interact with the data. The diagram in *Figure 4.1* represents the fundamental tasks of the API team. Incorporating a mature API product mindset influences the entire *digital value chain*, as depicted in the diagram. Care needs to be taken when building APIs so that end users have consistent access to record systems and other backend data:

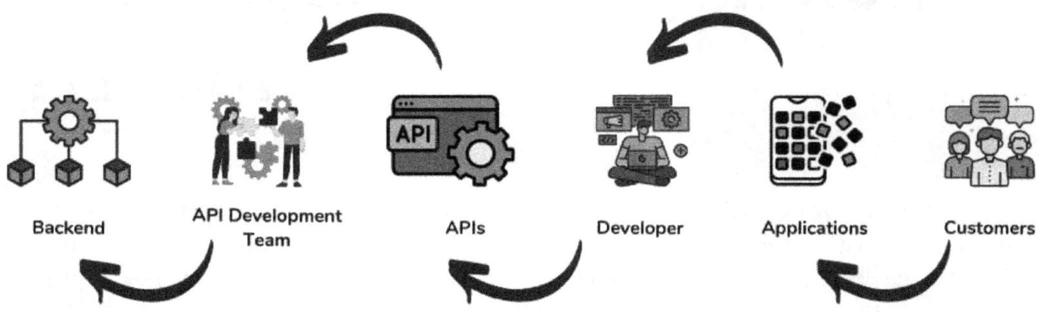

Figure 4.1 – APIs in the digital value chain

This chain points out the primary users of the API: our **Developer** and **Customers** – the end users of the application. The experience of both these types of users is influenced by our **API Development Team**.

Developers are often concerned with the ease of use, reliability, robustness, and scalability of APIs so that they feel confident when using them to build their user experience. End users, for whom developers construct apps and experiences, are often not aware of the APIs that work in the background of the applications they use. For APIs, the main customer is the developer who uses these APIs to build applications for end users. Even though APIs are two steps removed from the end user's experience, they influence the application's performance, features, and capabilities, and, in turn, the overall user experience.

Building a **Minimal Viable Product** (**MVP**), curating a world-class developer experience, and championing the commercial value of APIs are all important goals for the API team to pursue as they work toward establishing a sustainable value chain.

Leading with the API strategy

Designing and releasing APIs for long-term value at scale and continuously improving them to meet increasing customer needs are the hallmarks of an API product approach. In contrast, when APIs are treated as standalone initiatives or as parts of larger projects, they tend to be less effective in terms of their scope, durability, and accessibility.

Developing APIs with a product-oriented approach means treating them as such during the design phase. The ease with which an API can be utilized in the future depends on how well it is built to be consumed by developers. However, if the API is only intended to be used within the scope of a single project, its author may fail to take into account critical elements such as documentation, uniform design standards, versioning, security, and extensibility.

Instead, API designers that think like product developers will put future sustainability and extensibility first. The importance of this distinction should be obvious given that few things unify company leaders like the dread of being trapped into a certain use case, strategy, or business model does.

Setting the short-term and long-term strategy for API products tends to be tricky because there aren't always industry standards to follow. Even in the face of competition, nearly every company does things in its own way to address its unique challenges. This is both a blessing and a curse because while it allows you to shape the product, it also gives you limited reference material. As a product manager for an API product, understanding the business domain, customer use cases, pain points, and opportunities, along with having a technical understanding, is extremely important in being able to build and manage API products.

As with any product, the first version of an API can be very simple, but as the product matures, the challenges can get more complicated. In addition to growth and retention, monitoring churn patterns may become crucial to defining your product strategy. The API life cycle is closely tied to an API's maturity. At each step of API maturity, the stakeholders, customer needs, and expectations change.

The API development team

API development teams can be structured in a variety of ways depending on the size and needs of the organization. However, some common team structures for API development teams include the following:

- **Cross-functional teams**: This structure brings together developers, designers, product managers, and QA engineers in a single team. This approach allows for collaboration and faster development times, as all the necessary expertise is in one place.
- **DevOps teams**: This structure combines development and operations teams to enable the faster delivery of new features and improvements. This approach is particularly suitable for organizations that use the Agile development methodology and need to quickly iterate and deliver new features.
- **Centralized teams**: With this structure, the API development team is centralized within the organization and is responsible for developing and maintaining all of the company's APIs. This structure is suitable for organizations that have a large number of APIs and need to ensure consistency and quality across all of them.

- **Decentralized teams**: With this structure, the API development team is decentralized, and each department or business unit within the organization has its own API development team. This structure is suitable for organizations that have a large number of APIs and need to ensure that each department or business unit can respond quickly to its specific needs.
- **Outsourced teams**: With this structure, the API development team is outsourced to a third-party vendor. This approach is suitable for organizations that do not have the resources or expertise to develop and maintain APIs in-house.

The structure of the API development team will depend on the size, needs, and goals of the organization in question. The key goal should be to create a structure that allows the team to work efficiently and effectively and deliver value to the business.

In small organizations, API development teams are often small and cross-functional. The team might include a few developers, a designer, and a product manager. This structure is suitable for small organizations because it allows for flexibility and fast decision-making.

In medium-sized organizations, API development teams are often larger and may be divided into sub-teams based on specific areas of expertise. For example, a larger development team may include a frontend team, a backend team, and a QA team. This structure is suitable for medium-sized organizations because it allows for scalability and the ability to handle a larger volume of work.

In large organizations, API development teams can be very large and may be divided into multiple teams based on different business units or products. These teams may also have specific areas of expertise, such as security, performance, or scalability. This structure is suitable for large organizations because it allows for a high degree of specialization and the ability to handle a large volume of work.

In large-enterprise organizations, API development teams are often centralized and responsible for developing and maintaining all of the company's APIs. This structure is suitable for large organizations because it allows for consistency and quality across all APIs. Additionally, the team may be divided into sub-teams based on specific areas of expertise, such as security, performance, scalability, integration, and so on. As a company matures on its API journey, all product teams will have an API focus and develop API expertise. Having a centralized structure won't scale – unless all teams and product managers have an API mindset, APIs will be relegated to just being tech interfaces (as opposed to being treated as products).

API development teams typically consist of a variety of roles, each with its own specific responsibilities. Here are a few examples of some common roles on an API development team:

- **Product managers**: The product manager is responsible for defining the overall vision and strategy for the API product. This includes identifying target audiences, creating product roadmaps, and determining which features to include in the API. They are responsible for the business side of the API development.

- **Developers**: Developers are responsible for writing the code that makes up the API. They work closely with the product manager and other members of the development team to ensure that the API meets the needs of the target audience and is built to the highest standards. They are responsible for the technical side of the API development.
- **Test engineers**: Test engineers are responsible for ensuring that the API functions correctly and is bug-free. They work closely with the development team to create test plans and test cases, and then execute those tests to ensure that the API meets the required quality standards.
- **Architects**: Architects are responsible for designing the overall structure and layout of the API. This includes determining the overall architecture of the API, designing the database schema, and creating the API's endpoint structure. They work closely with the development team to ensure that the API is designed to be scalable and secure.
- **DevOps engineers**: DevOps engineers are responsible for managing the infrastructure and tooling that support the API development process. They work closely with the development team to ensure that the API is built and deployed efficiently and that the API can handle traffic and scale as needed.
- **Business analysts**: Business analysts are responsible for gathering and analyzing data to understand the needs of the target audience and the current market trends. They work closely with the product manager to identify opportunities for new features and improvements to the API.
- **Technical writers**: Technical writers are responsible for creating documentation for the API, including API reference materials, developer guides, and tutorials. They work closely with the development team to ensure that the documentation is accurate, up to date, and easy to understand.
- **API evangelists**: API evangelists represent the API's users. They manage the developer portals and ensure that developers have access to product documentation and SDKs. Developer outreach is their responsibility. This outreach may include answering queries, conducting hackathons, and bringing developer input to the API team to shape the product roadmap. They also play a role in marketing the APIs via SEO, targeted ads, and other techniques. Evangelists can also attract partners to a company's APIs, extending its network of users.

These are just a few examples of common roles on an API development team. Depending on the size and needs of the organization, there may be other roles as well. For example, in highly regulated areas, compliance and legal teams are an important part of the API development team and advise on data policy and privileges. The key goal is to have the right mix of skills and expertise to develop, test, deploy, maintain, and improve the API product.

These functions can often be spread across the organization and handled by either small or large teams. As you build APIs, you need to create awareness and get buy-in from all these teams. Most importantly, you must bring them together to help build and support the API.

Managing an API as a product

An API as a product involves treating the API as a standalone product that is marketed, sold, and supported for customer use. This includes tasks such as defining the API's value proposition, creating documentation and tutorials, setting the pricing and usage limits, monitoring the usage and performance, and responding to customer feedback and support requests. It also involves continuously updating and improving the API based on customer feedback and market trends to ensure it remains valuable to customers. In other words, managing an API as a product is similar to managing any other product, but with a focus on the specific needs and considerations of an API.

Organizations that already have APIs can transition to managing them as products by implementing the following steps:

1. **Define the value proposition of the API**: Understand the unique value that the API provides to customers and how it addresses specific business needs or solves specific problems.
2. **Create comprehensive documentation and tutorials**: Make it easy for customers to understand how to use the API and what they can do with it.
3. **Set clear pricing and usage limits**: Establish a pricing model that aligns with the value the API provides and set usage limits that are fair and transparent.
4. **Monitor usage and performance**: Use analytics and monitoring tools to track how the API is used and identify any performance issues that need to be addressed.
5. **Respond to customer feedback and support requests**: Establish a process for handling customer feedback and support requests, and make sure the team responsible for the API is responsive and attentive to customer needs.
6. **Continuously update and improve the API**: Use customer feedback and market trends to continuously update and improve the API to ensure it remains valuable to customers.
7. **Communicate effectively with customers**: Communicate the value proposition, usage, changes, and updates effectively to customers and make sure that they are aware of any changes.

By following these steps, organizations can begin to manage their APIs as products and create a more sustainable, revenue-generating business model around their APIs. Businesses may encounter certain difficulties as they transition from building APIs for use within projects to building and managing API products. In many organizations, different project teams are free to choose their own software libraries and other implementation details, which can lead to unexpected complexity and cost if the APIs are exposed in a way that generates dependencies.

Consider the case of an undocumented API whose payload contains a data element labeled *location* but what it actually delivers is a shipping address. This naming strategy may have made sense to the original developer within the context of their specific project, but subsequently, future developers attempting to use the API may become confused due to a lack of documentation and ambiguous nomenclature.

To adopt the API product approach, you need to treat your API like a real product. A product manager, as with any other product, should be in charge of it rather than being a project manager whose job is to fulfill a set of requirements. The product manager is accountable for learning about the product's target audience and turning their needs into product specifications and an iterative development plan. To carry out these duties successfully, the product manager must have the support of key business and technical executives on the objectives and benefits of the API initiative. To better meet the SLAs, gain insight, and guarantee that the product satisfies customer expectations, they should also have access to API management tools that reveal how and by whom the API is being utilized.

The most successful API programs are usually led by groups whose primary focus is on creating APIs as products with the express goal of increasing developer efficiency. To keep up with the ever-changing demands of their businesses, many companies now handle APIs in the same manner as physical products, complete with their own lifespans and roadmaps. APIs can be powerful business accelerators, but only if companies approach them as products rather than projects.

The API proposal

In a small organization, the API team can usually build APIs independently, but in most larger organizations, an API governance team will approve any APIs being built in the organization. Before you start building an API, you must put together a product specification document that talks about what the API is required to do. This is a simple document that maps out the API workflow and narrates the story of why this API needs to be built.

This API proposal document should map out the workflow of the API, along with various user stories. User stories are written in the format shown in the following figure:

As a {user persona}
I want to {desired goal}
so that I can {benefit}.

Figure 4.2 – User story format

This format is widely used across all types of products to the goals that need to be achieved by it and communicate the actions that need to be taken in the process.

For example, if you propose that your team builds a shipment tracking API for an e-commerce company, you can start to imagine user stories such as the following:

- As a customer, when I place an order through the website, I get an email notification with tracking information as soon as my order has been shipped so that I am informed.
- As a customer, when my order is shipped, I can see the status of the order and shipment information in my account so that I can log into my account and review my orders.

- As a customer, when my shipment is delayed, I get an email notifying me that my shipment has been delayed and I am informed of when to expect the delivery of my order.

These stories will allow the development team to understand the functionality expectations for the API that you're proposing to build. In this example, with three simple user stories, you can see that this API will be used to send email messages to customers. This API is also expected to be integrated into the site, where users can log in and review shipment statuses, along with order statuses. The third story talks about designing workflows based on specific shipment statuses.

API stories must cover the purpose of the API and explain why it should exist. You should also provide a list of all the data properties (inputs and outputs) that the API needs to do its job, all of the things that the API should be able to do, any rules that the API should follow, and any tasks that the API is in charge of completing.

If you think further about this example, there will be scenarios in which users will decide to cancel their order while the shipment is in flight. This will lead to a workflow of handling returns, where the customer needs to be issued a refund when the customer has returned the shipment after receiving it. This will also require the customer to receive a returns label.

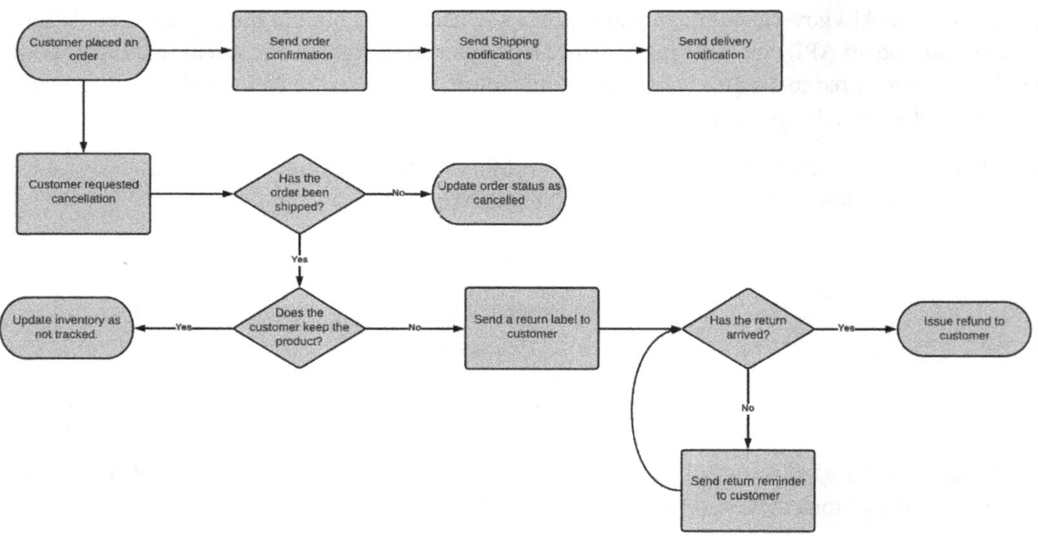

Figure 4.3 – Order processing workflow showing order notification scenarios

It is also very useful to draw a simple diagram to illustrate the workflow so that the requirements can be understood based on the various states of orders and shipments. These diagrams can also help teams walk through customer scenarios and identify opportunities for improving the customer experience. Based on this discovery and analysis of the user flow, you can make decisions for the design of APIs.

Designing APIs

In most organizations, the API program will be ongoing and involve numerous departments and employees, as well as a wide range of tools and resources. Maintaining a consistent design approach or method is one way to provide continuity amidst the chaos of constant change. If you train all of your developers to use the same design process, you ensure uniformity without stifling their individuality.

Additionally, a solid design process does not rely on a single technology stack or collection of tools. As a result, your API program can evolve over time depending on the power of consistent design methods, rather than a specific API format, protocol, or another technical component. In fact, having a solid design process in place makes it less of a hassle to update tools without introducing incompatible new implementations. Therefore, we should design tools to help make API software consistent and compatible.

When designing APIs, you will partner closely with your engineering leads, who will review the user stories you presented in your proposal and collaborate with you on putting together a prototype of the API you are proposing to build.

Ensure that the APIs you create are in line with user expectations and can sustain a sustainable business model. It's not just the end users but also the developers that need to be taken into account when creating APIs. APIs that effectively address a problem can only be developed after careful consideration of the target audience. When building an API, it's important to keep in mind who will use it. Each individual is unique and hence has a unique perspective, set of abilities, and set of priorities.

It is also crucial to make APIs that help an organization achieve its goals. A solid starting point for designing and developing APIs is to ensure that they will be used and will help address a real business problem. It is not a good idea to build APIs just to have them. It is also a bad idea to build APIs that nobody has requested or requires. I've seen API teams work hard to secure funds from upper management to develop and release APIs. However, upon release, it is observed, to everyone's dismay, that the APIs aren't suited for the business case they were supposed to address.

In order to create APIs that *fit people's needs* and back up a *viable business strategy*, you can follow a four-step design method as follows:

1. Review all of the names of the data and action elements and, if necessary, change them to names that your API program already knows.
2. Turn your written notes about how work gets done and how things work internally into something that people can see and understand.
3. Make a document that developers can use as a starting point for implementing the API. This document should be machine-readable.
4. Write basic reference materials for both service developers and people who use the service (those using the API in the future).

These steps will prepare you to build MVPs for your APIs that can be reviewed by the architects within your company and explored to validate the foundational hypotheses that led to their creation. In the following section, you will read more about the importance of building MVPs.

Starting with an MVP

Companies that are focused on their products typically believe that releasing an MVP to the market (or, in the case of an internal API, to production) as soon as possible and then continuously iterating based on developer feedback is the ideal strategy. By shifting to a product-first mentality, businesses can rapidly test out new ideas across the board and adjust their spending based on what works and what doesn't.

Since an API management layer can help separate the quick development of apps and the onboarding of new partners from the slower, more rigid processes of the company, an API-driven organization may be more agile than its more traditional counterpart.

However, developing an API isn't enough to make a company more agile. Many corporations lack the infrastructure to accept internet-scale partners into their annual project plans. The sales, marketing, and finance departments, among others, may have developed their own rhythms over time. So that developers can quickly open up new business options, it is important to envision iterating API solutions that include build and release procedures, version control, and measurement and reporting.

All of these departments should be brought together by the speed that APIs offer and the commercial potential they can open up, which is the responsibility of the API team. In one direction, the team must coordinate with internal and external developers and business divisions, and in the other, it must coordinate with backend IT teams. By rapidly exposing concepts to the market, gathering data, and iterating, it should produce a continual flow of new product ideas.

As you develop the MVP of your API, you can engage with the customers who are most passionate about this API and get their feedback. For internal APIs, these could be internal teams. For external facing APIs, these could be solution architects, sales engineers, or even technical sales staff who are familiar with the customers most likely to use these APIs. You can also partner with user research groups to run research studies to get feedback from developers directly on how they would use these APIs.

Constant engagement with customers to get feedback allows you to learn quickly about the gaps in the product, as well as the changing needs of your customers, and enables you to iteratively improve APIs. In the next section, you will learn more about an iterative way of building APIs.

Building and releasing APIs in an iterative way

A product-focused API provider can gather insights to groom a backlog of updates and decide whether to mature or shut down features or products through an iterative process of creating hypotheses, testing them, and unpacking discoveries. API providers have the responsibility to keep up with changing trends and the demands of developers so that they continue to meet the needs of their users and satisfy their curiosity.

On the other hand, projects normally have a defined beginning and finish time. In most cases, organizations that want to improve upon the results of a project must start from scratch by assembling a new team, securing additional money, and relaunching the program. To put it simply, this is a rather inconvenient method of doing things. As a result of having an MVP, continuing operational processes, and a product manager responsible for assessing and pushing the product's performance from the start, businesses that adopt a product mindset are able to bring it to market and iterate more quickly.

APIs are used by customers to build their applications and changes to APIs can break their applications. This is why APIs evolve iteratively. To ensure that customers are not handed a half-baked process, an iterative release flow enables experimentation and user feedback before APIs are published to a wider audience. APIs are distributed and released to customers based on various factors, such as the maturity of the APIs and the target customer for which they are designed. The following figure shows how every development iteration of an API goes through the release process based on whether it is targeted for internal release, partner consumption, or public consumption. The diagram also shows how, based on the maturity of the API, you can identify the supporting API experience components to enable customers to use the APIs:

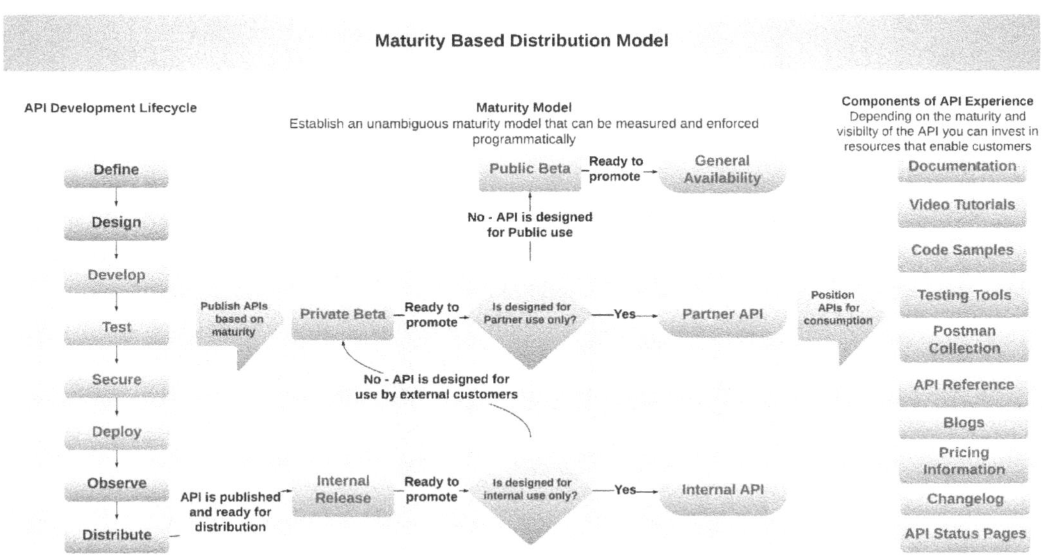

Figure 4.4 – Maturity-based distribution model for API products

As you add new features and functionality to your APIs, they are tested with internal customers first. These could be sales teams, solution architects, or other developers within the organization. This also helps to create an awareness of the new functionality among the stakeholders and helps you gather feedback. If these APIs are only meant for internal use, there might be a limited set of stakeholders, and their approval might be sufficient to complete the work on the APIs.

If the APIs are built for external consumption, however, using input from internal consumers, you can establish whether the product meets the customers' needs to a reasonable degree. If so, you can move the APIs to a private beta stage in which you allow only a few customers to test out the product. Based on closer collaboration with these customers, you can make improvements to the APIs. The goal of this process is to limit the exposure of a less mature product to a handful of customers who are aware of the beta status so that they can provide feedback without building dependencies on the APIs. In the beta stage, the APIs are subject to change, and setting expectations with the customers enables them to engage with the team for better results.

For example, APIs that are built for specific customers or partners might be limited to these partners and customers for feedback. Once these partners and customers approve of the product, the APIs may be considered complete. Based on the guidance of the API governance standards within your organization, you might have to make this documentation available to these partners and customers along with code samples, API references, API status pages, and testing tools.

For public APIs, the APIs transition from private beta to the public beta stage, where they are made available to all customers with the label of beta to set expectations on the maturity of the APIs. As you start to expose the APIs publicly, you will want to collaborate with the developer relations team to publish the necessary blogs, tutorials, and social media posts to raise awareness around the product, alongside publishing documentation, code samples, API references, and API status pages and setting up testing tools.

The API experience components that support APIs are aligned with the maturity of the APIs, and the API governance team should provide guidance on the API experience components that are needed for a generally available API.

The release process of APIs can be complex and span over long periods. You need to ensure that you have the necessary resources and personnel available to manage and improve API programs over time. A product manager's ability to use metrics and user feedback to advocate for an API program's worth is crucial to securing this kind of funding and stakeholder alignment. You must set both short-term and long-term strategies for the API program so that you can make the case for continued investment in the API program. You will learn about the importance of short-term and long-term roadmaps in the next section.

Building long-term and short-term roadmaps

Building an effective MVP allows you to get your iterative process of API development started. Although in the most basic sense, charging for access to an API is the most common method of making money off of it, there are actually quite a few different motivations for doing so.

APIs are granular, time-bound, high-value endeavors. Teams can work in parallel, which is beneficial to the company's growth when different parts of the business are decoupled and packaged as clear, specialized APIs. APIs allow the company to reach a wider audience, which can lead to more sales, better customer service, and the discovery of new avenues for collaboration with other businesses in the industry.

In an API ecosystem, internal and external products and services can freely share and receive data from one another. The popularity of APIs as a measurement of consumer value is a sure bet. By themselves, API implementations help you better understand which aspects of your business are most important. This is useful for rethinking your current technological solutions and fine-tuning the implementation specifics.

In the first few iterations of the APIs, you may only engage with a few customers, but as you start to get a wider set of customers, the requests from customers will grow. You can then start to think about reviewing the major trends in requirements from your existing customers and compare them with the business goals and the target audience you are trying to reach to find the best alignment with set priorities.

In the short term, with a small set of users, your team could focus on expanding the functionality of the APIs so that more customers become interested in using the product. However, as the customer base grows and their usage increases, scaling will become a priority to enable your APIs to support the usage needs of a larger customer base. If you are operating in a highly regulated domain, such as finance or healthcare, making architectural improvements to be more compliant might also be a key factor in unlocking growth in your segment. All these aspects will translate into a dense and impactful backlog that you can review with your team regularly and validate with user research to establish a short-term and long-term roadmap that truly reflects the trajectory of your product.

Summary

The API development team works together to build APIs that serve the customers; developers build applications using these APIs that serve the end customers. Even though the API development team is a few steps removed from the end customers, as we saw in the digital value chain, their work is crucial to enabling developers to build applications that impact millions of people across the world. Building APIs that lay the foundation for building innovative applications also unlocks innovation for developers across the globe.

You can build APIs iteratively and methodically to ensure that you are keeping up with the fast-changing needs of the customer base while keeping an eye on the business goals at all times. Working across the organization to bring together security teams, development teams, testers, developer evangelists, and sales and support teams to deliver APIs that are intuitive, coherent, and consistent is the most rewarding aspect of the role of the API product manager.

In the following chapter, you will learn about identifying and quantifying target markets for APIs and building sales, marketing, product-led, and community-led growth strategies to help increase the usage of your APIs.

5
Growth for API Products

Early in the product life cycle, we spent time validating the product demand. Once we have established a target audience and a market need for a product, we must find ways to reach this audience and capture as much of the market as possible. Growing the product's adoption involves creating a targeted marketing strategy to raise awareness about the product in the target audience, as well as developing a sales strategy that drives revenue.

Because APIs are mostly used by developers who work alone or as part of a team at an organization, they may need to be grown differently than consumer products. How well we can show customers the usefulness of an API product is very important for its growth.

In this chapter, we'll look at some of the most important ideas and strategies that can help API products grow:

- Understanding what growth means for APIs by doing the following:
- Product-led growth
- Community-driven growth
- Low-code and no-code integrations

Understanding what growth means for APIs

APIs not only serve as a way for a company to offer a capability to external customers but are also a big part of its internal infrastructure. Internal APIs allow teams to build infrastructure capabilities that other groups within an organization can use. Consequently, most APIs are made to be internal or for partner usage only and are not public-facing.

The definition of growth for an API product is deeply aligned with whether the APIs are for internal, private, or for public usage, as this will determine who the customer is and the tools they have available to discover and start using a given API.

In the following sections, you will learn about the growth potential of each of the aforementioned APIs.

Internal APIs

For internal APIs, the internal tooling makes APIs readily available to other teams. Usually, an API architecture or governance team would establish best practices for how APIs are built and published across an organization. Some tools, such as Postman, are utilized to share an API in the form of collections to help developers get started with APIs quickly. Postman allows a GUI-based interface where users can easily plug in their API credentials and get started. Tools such as Postman can often generate API reference documentation from API specifications to aid the adoption of APIs.

The potential for growth of internal APIs is limited by the size of an organization because that is the size of the user base that can potentially use this API.

An e-commerce company's API for tracking shipments is an example of an internal API. This API can be helpful for building functionality, such as sending users shipment notifications as their order is shipped. This could be through emails or SMS alerts. Such functionality also needs to be surfaced to customer support teams so that support teams can answer questions about shipment delivery when customers reach out. If the team building the API doesn't make it known that such functionality already exists, then other teams would build their own separate version, which would result in duplication of work, a waste of engineering effort, and a loss in terms of delay. To avoid this scenario, internal processes are created to aid in the discovery and documentation of internal APIs.

Internal APIs, although used within an organization, must be treated like products in their own right and marketed effectively to raise awareness and adoption within the organization. As internal APIs get consumed widely within the organization, it shares the cost of development and operation, ultimately reducing redundancies.

Partner APIs

A partner API is a type of external-facing API that is usually made for a specific customer or partner. Because of the close collaboration with the customer, there is more specificity to how these APIs are built.

You can go from an MVP to a more mature partner API reasonably quickly because the customer is closely engaged and driving the requirements. In terms of growth, partner APIs would grow in their usage, as the partner could increase usage on their platform and among their customer base. The partner's growth sets the limits for the growth in the usage of partner APIs.

Partner APIs often drive a sizable portion of business, and investing in the quality of these APIs can help get additional partners in the future.

Public APIs

Public APIs are made available to anyone and everyone, and the market for such APIs is often global, depending on the nature of the functionality being offered. The growth of public APIs is based on how easy it is to discover them and integrate with them. However, the means to make it easier for customers to discover and integrate with public APIs tend to be more numerous. A good example of public APIs is SendGrid's email APIs, which enable a vast set of email interactions using APIs. Complex marketing and customer support operations are built using these APIs, and products such as Shopify and Klaviyo use SendGrid APIs to offer advanced email functionality for their users.

Public APIs are designed for a **business-to-business** (**B2B**) audience, and the ideal customers would be using these APIs at scale. Growth for a public API is very focused on large-scale customers and their ability to provide a reliable, robust, and scalable experience.

You are now aware of the types of APIs based on their usage and how to assess their growth potential. You were also introduced to the target audience for each API type. In the following section, you will learn about the methodologies behind identifying target audiences.

Identifying the target audience

Finding the right target audience for your product is a crucial step in the product development process, as it helps to ensure that your product or service is well aligned with the needs and wants of your customers. Here are a few ways to identify your target audience:

- **Conduct market research**: Conduct surveys and interviews with potential customers to understand their needs and pain points. For example, you might ask developers about the types of data and functionality they need from an API.
- **Analyze your competition**: Look at the target audience of similar API products and consider whether those audiences align with your own. For example, you might find that your competitors' APIs are geared toward small businesses, while your API is intended for larger enterprises.
- **Use demographic data**: Look at data such as the industries and company sizes of potential customers to understand your target audience better. For example, you might find that your API is most relevant for customers in the finance and healthcare industries.
- **Define your ideal customer**: Create a detailed profile of your ideal customer, including their job role, industry, company size, and pain points. For example, your ideal customer might be a senior developer at a large enterprise with a need for data security and privacy.
- **Test your product or service with a small group of potential customers**: Reach out to a small group of developers and ask them to test your API. Collect feedback on the functionality and ease of use of the API to help identify any issues that need to be addressed.

- **Use social media listening and analytics tools**: Use tools such as Google Alerts, SocialMention, and Hootsuite Insights to understand what developers are talking about in relation to your API.
- **Consider the customer journey**: Map out the journey of a developer from the discovery of your API to the integration and maintenance of it, and identify the pain points and areas where the developer needs help.
- **Use segmentation, personas, and journey mapping**: Segment your customers based on their industry, company size, job role, and so on. Create personas that represent each segment and map the journey of each persona to understand their needs and pain points.

By using a combination of these methods, you can gain a better understanding of your target audience and ensure that your product or service is well-aligned with their needs and wants. Additionally, regularly monitoring your target audience can help you adapt your product or service to changing market trends and customer needs.

Once you have your target audience identified, you can find a way to understand the potential market size for the product. To establish the market size for which a product is targeted, we use the methodology of calculating the total addressable market for the product. The **Total Addressable Market** (**TAM**), sometimes called the **total available market**, is the total amount of money that a product or service could make if it had 100% of the market. It helps a person or company decide how much time and money they should put into a new business line.

In most large organizations, the market research teams help with the calculation of TAM. However, in smaller organizations, you might have to establish TAM on your own to be able to make product decisions. For example, if you are building a shopping app specifically for senior citizens, you would calculate TAM based on the number of seniors using smartphones, how much they are spending on online shopping, which geographies this app would work in, and so on. Thus, the total market size is given as a dollar amount that shows how much money this shopping app could make. This paints a somewhat high-level picture of TAM, but there are some methods involved in calculating it in a professional context. There are three primary methods to calculate TAM:

- Top-down
- Bottom-up
- The theory of value

Let's look at each of them in detail in the following section.

Methods to calculate TAM

The top-down TAM model

The top-down approach to calculating TAM involves starting with the overall size of the market and then segmenting it to determine the market share you can address. Most of the time, you can find the top-down calculation in research publications by third parties that give industry insights. The top-down TAM model is used to gain a high-level evaluation of the market so that you can evaluate whether the market opportunity is worth entering.

The following diagram shows an example of a calculation for the data backup market in North America, using a top-down approach.

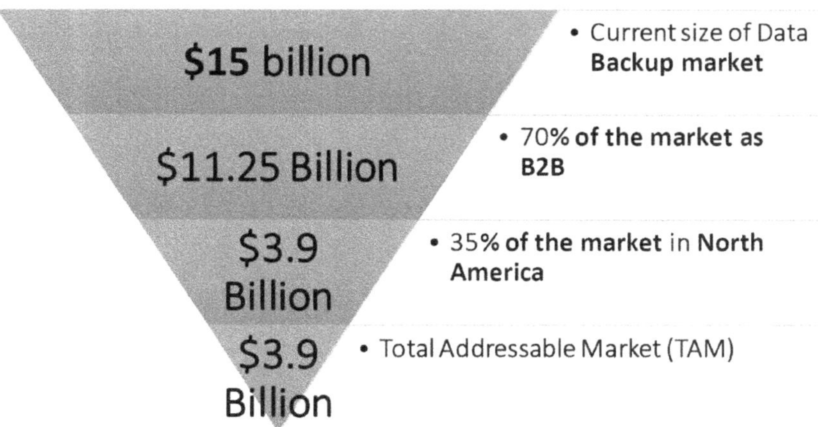

Figure 5.1 – A top-down TAM calculation example for the data backup market

In this example, you can see how we start with the total market size of current data backup markets around the world and then start narrowing down, based on the B2B market share and geography, to arrive at a TAM figure of $3.9 billion. This approach provides a quick way of assessing the market, but the difficulty of this approach lies in being able to evaluate segments, which is addressed by the bottom-up approach in the next section.

The bottom-up TAM model

The bottom-up method for figuring out TAM starts with focusing on the main market you've already found. Then, you add any nearby market segments to get an estimate of TAM. This method is more reliable than a top-down TAM calculation because it is more specific in how it finds and groups market segments for the calculation.

The bottom-up method of calculating TAM gives you more time and space to think about why you want to include some customer segments and not others. You can also use data from several sources, such as research or surveys, for each segment to form more concrete reasoning around your market sizing strategy.

The following diagram shows an example of a TAM calculation using a bottom-up approach for a B2B SaaS product.

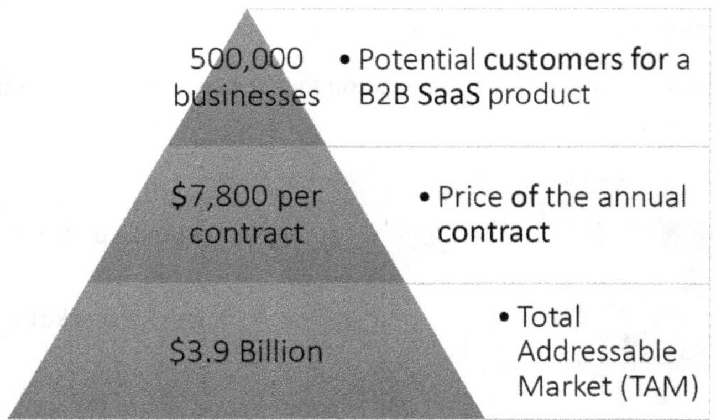

Figure 5.2 – An example of a bottom-up TAM calculation for a B2B SaaS product

In this example, you start with the number of potential customers and multiply it by the price of annual contracts currently seen in the market, arriving at a TAM value of $3.9 billion.

Although the bottom-up TAM calculation approach is more reliable and granular than the top-down approach, this approach only works for existing markets with a significant user base. This method also assumes that the starting point of your calculation is the best place to start from a strategic point of view, which may or may not be true.

Theory of value

Value theory addresses the gap in a bottom-up TAM calculation by providing a way to calculate TAM for new products that do not already exist on the market. The value theory TAM calculation is based on estimating how much value your customers perceive your product to deliver and how that is reflected in product pricing.

Uber uses Google Maps APIs to deliver the mapping functionality in their applications. Building an in-house mapping functionality could cost Uber years of development and millions in investment, not to increase their time to market. By using Google Maps APIs, Uber can focus on improving the user experience and pay Google for mapping features at a much lower cost. This data can be used to calculate TAM for Google Maps APIs as more and more businesses and platforms such as Uber choose to build with their APIs, rather than building their own mapping functionality.

Using value theory in combination with bottom-up TAM calculations can help you get a very granular and insightful view of your potential market and be an effective input to your product strategy decisions. You should not overestimate the size of the market because this could lead to markets that have less room for growth. For a product opportunity to be worthwhile, the best market is one that has room to grow. Based on the TAM calculations, you can establish the marketing, sales, and pricing strategies for the product that you will learn about in the following sections.

Marketing strategy

A marketing strategy is a detailed plan that typically spans several years. It establishes a clear understanding of customer requirements. A sound marketing strategy is critical for gaining and maintaining a competitive advantage. At its heart, the strategy helps you figure out who your customers are and how to reach out to them.

The marketing strategy also explains how you decide where to put your product. It helps establish a roadmap of products, their associated variants, and perhaps even features. As is commonly understood, the marketing strategy provides guidelines for advertising, partnerships, and promotions.

The marketing strategy for APIs is designed in very much the same way as any other product, but it does come with its own set of challenges, based on its unique technical nature. Let's now look at the components that constitute the marketing strategy:

- **Target audience**: A target market is a group of people who have some things in common and who a business thinks might be interested in its products. Finding the target market helps a company make decisions about how to design, package, and sell its product. Target audiences for your APIs could be developers, small businesses, or large enterprises across various industries and domains. It's important to recognize the target audience for your products clearly as you shape your marketing strategy.
- **Value proposition**: A *value proposition* is a short sentence that tells customers why they should buy your product or service. It makes it clear to customers what they get from doing business with you. Every value proposition should address a customer's problem and show how your business can solve it. For example, the Google Maps Platform's value proposition is presented as "create real-world, real-time experiences with the latest Maps, Routes, and Places features from Google Maps Platform. Built by the Google team for developers everywhere." A good value proposition may talk about what makes you different from competitors, but it should always focus on how customers see your value.
- **Product mix**: The number of product lines and individual products or services a company sells is its *product mix*. This is also called a product selection or a product portfolio. Each company has a different mix of products; for a large company, this might be a number of products, while for start-ups, this might be a single product or service. For example, Twilio provides APIs for SMS, voice, email, and so on. These product offerings are designed to complement each other in a product mix that can be bundled together to create advanced communication applications.

- **Brand messaging**: Brand messaging is how your brand's verbal and nonverbal messages share its unique value proposition and personality. Your message can make them feel inspired and want to buy your product. It's the way your customers can connect with your brand. With APIs, you are messaging developers who would integrate with your APIs to build applications. In the clearest possible way, you should design the brand messaging so that it drives the value of your APIs to these developers in terms of value proposition and ease of use.
- **Promotional strategy**: A promotion strategy is essential for positioning your brand on the market and letting people know about the products and services you offer, along with how they could benefit from choosing you. A promotion strategy is the plan and methods you use in your marketing plan to make more people want to buy your product or service. Promotional strategies are an important part of the marketing mix (product, price, placement, and promotion) and are based on the target audience, budget, and plan of action. Developer relations teams work closely with the marketing teams in the promotional strategy for your APIs across social media, blogs, white papers, and so on to make sure the right audience is able to receive the information about your APIs in a medium of their choice. These teams also work to build a developer community around your APIs for a sustained cycle of promotion for them.
- **Content marketing**: This is a strategy to gain relevance through the proliferation of content about your product via various mediums, such as print, and video. The objective is to reach as many potential customers as possible and make them aware of your product or company. Both traditional mediums, such as magazines, as well as modern ones, such as social media, should be considered. Most attention should be given to a medium that is preferred by your target audience. Developers often learn about APIs from platforms such as YouTube, Udemy, and technology blogs. These platforms can be great mediums for getting developers to discover and try out your APIs.

Create a marketing strategy that not only makes sure that your products meet the needs of your customers but also tells them about them. The strategy guides you on how to approach your customers and stay laser-focused on targeting them with the right products, effectively keeping your spending under control.

You will work closely with the marketing teams to understand how to market your products. A well-researched marketing strategy is expected to align with and complement your sales and pricing strategies, which you will learn about in the next sections.

Pricing strategy

Depending on how much it costs, a good product can do well or badly on the market. Products that are priced too high will be beaten by competition that prices them fairly. On the flip side, pricing low may bring higher sales but may come at a narrow profit margin or even a loss.

The price of an item also affects how much people think it's worth and how much your brand is worth. It can help paint a picture of how desirable, useful, popular, or good a product is. Underpricing a product can be a good way to compete, but if it's the wrong product, it can also make people think the product isn't worth as much. Also, if you depend on partners to sell your products, they may give preference to competitors with higher prices because they bring in more profit, even if the customers get less for their money. In this case, dropping your prices can actually make you less competitive, not more.

The pricing strategy needs to be changed, based on factors such as where customers live, how mature the product category is, and the state of the economy as a whole. Thus, there are several product pricing strategies; however, in the following section, you will learn about some key ones that are of interest to API products:

- **Freemium pricing**: Freemium pricing is utilized by several **software as a service** (**SaaS**) providers, such as Twilio, Slack, and Dropbox. The freemium pricing model is often used along with a more traditional paid tier, making it a tiered price structure. The free tier is often restricted in terms of features, usage limits, or use cases. This encourages users to pay for a higher tier once they reach a certain usage threshold. The freemium business model streamlines the onboarding process for new users of your product. Due to the low barriers to entry, products are able to realize growth as word of mouth spreads from satisfied customers, ultimately reducing the **customer acquisition cost** (**CAC**). Twilio Email APIs have a free tier that allows you to send 100 emails per day. This allows customers to be able to build a proof-of-concept email application without any cost and then scale their usage to one of the multiple paid tiers, based on their needs.
- **Usage-based pricing or pay-as-you-go pricing**: Software providers that focus on infrastructure and platforms (such as Amazon Web Services) frequently adopt a usage-based pricing model, in which customers are billed based on API calls, transactions executed, or bytes of data used. This pricing technique, also known as the pay-as-you-go model, closely correlates the price of a SaaS product with its usage – the more you use a service, the higher your bill will be, and the less you use it, the lower your bill will be. Usage-based pricing allows users to tie price to consumption so that they pay less in slower months. This model allows users to take comfort in being able to scale their costs based on their usage, without any upfront costs. However, because your revenue is tied to customer usage that might have unpredictable seasonality, this model makes it hard for you to predict revenue.
- **Volume discounts**: Following directly from usage-based pricing are volume discounts. API calls don't always translate to business transactions, and often, each business transaction can require multiple API calls, depending on the way applications are developed using your APIs. As customs scale their usage of your APIs, you can consider offering volume-based discounts to facilitate their growth. For example, Twilio SMS offers usage-based discounts, where the first 5 million messages are priced at $0.0079, while the next 20 million messages sent or received are priced at $0.0053, with the next 75 million messages sent or received being charged at $0.0032, and so on. This volume-based pricing model particularly works well for large-scale customers.

- **Feature-based pricing**: The feature-based pricing model is a type of usage-based pricing model that involves creating pricing tiers differentiated by the features available. This model is increasingly becoming popular with generative AI products, such as Jasper.ai and Twilio email. Jasper.ai is a generative AI product that writes short-form content, such as ads and product descriptions, and improves existing copy. You can purchase a different number of words per month through three different feature-based pricing tiers. These tiers offer different features, such as better support or a better editor experience. The business tier also offers features for team management and dedicated account managers.

The best time to price a product is at launch. As we've seen, excessive prices can drive people away unless they have the proper buzz or quality. Too low a price without proper marketing can give potential customers the wrong impression and a misunderstanding of the value of your product.

Explore the market to learn about similar items, how they're performing, what makes them popular or unpopular, and why buyers will pay for them. Understand your audience and why they want your goods. Describe your product's worth and why a customer would buy it. Focus groups can help you learn what customers find reasonable. Know your target market and the product's purpose. Then, determine the optimal price approach.

Based on your understanding of the target audience, you can now set the pricing strategy that matches your marketing strategy. You will be able to combine the pricing strategy and marketing strategy to establish a sales strategy, which you will learn about in the next section.

Sales strategy

A sales plan helps B2B sales teams meet sales goals and engage in direct selling. A sales plan determines who you'll sell to and how you'll accomplish it, which is vital to any business's revenue. Implementing a sales plan improves team performance, targeting, and a closed-won deals ratio.

In order to win in B2B sales, you must be strategic. It also requires developing a sales process so that your team can track won and lost deals methodically. Ultimately, better strategies generate more revenue. As you saw in the digital value chain in *Chapter 4*, APIs are products that are focused on B2B sales, and you will often work with a sales team as an API product manager.

There are two ways to sell something – **inbound sales** and **outbound sales**. You can choose one or the other, or you can do both.

Inbound marketing can be thought of as a sales strategy because marketing materials grab the attention of buyers. Through this process, they already know what your company sells, so they are warmer than outbound leads when they enter the sales funnel. The content is then used to teach, inform, and guide prospects through the process of making a decision.

Outbound lead generation is more about sending emails and making cold calls to prospects. These buyers haven't shown any interest and often don't know anything about your company, so it's down to your rep to convince them that your solution is a good fit.

Many different parts make up an excellent B2B sales strategy, but they all work together. Each piece of the strategy is linked to the others, creating a feedback loop that improves performance and makes the process more efficient.

An effective sales strategy is made up of these five parts:

- Setting up a sales process
- Implementing sales tactics
- Optimizing sales performance
- Establishing data-driven tracking
- Implementing sales tooling

Let us dive deeper into each of these.

Setting up a sales process

The first step of a sales strategy for B2B is to step back and look at where a business currently is. By looking at the analytics, you can set goals, metrics, and **key performance indicators** (**KPIs**) and plan how to achieve them.

Start by setting B2B sales goals. How much do you want to make every quarter and every year? Then, Custommake your plan based on that number.

Your KPIs and metrics for getting sales are the most critical part of your sales strategy. B2B sales metrics are the numbers you keep an eye on to ensure that you are on the right track. Your job is to figure out which metrics are the most important. Focus on the metrics you've chosen and work to improve them. If the numbers are going down, you need to do something.

Find the type of person in your target audience who will find your product the most useful; this is your **ideal customer profile** (**ICP**). Look at the size of their company, the quality of their data, and how they buy things. Then, explain how your product's features will help your target customers solve their problems. This is directly driven as a result of the TAM calculation.

The process of figuring out who your potential customers are, starting a conversation with them, and leading them to the next step in your SaaS sales strategy. The next step is to implement this strategy.

Implementing sales tactics

Tactics are how an effective sales strategy is put into action. You know how you want to sell, so now you must ensure that you have a process that enables successful sales. How will you get in touch with your ICP, and what will you say?

Create buyer personas to help you figure out which channels to use, and then create a sales process playbook that will help you get the desired results. User research teams are your partners in defining buyer personas. For example, if you think cold calling is the best way to reach your prospects, make sure your salespeople know how to talk about their problems and how your product can solve them. This will be the sales script or the pitch that sales teams are equipped with to be able to make cold calls.

Lead qualification is the process of dividing your customers into groups to find your best customers and the ones who will use your products the most. **Sales development representatives** (**SDRs**) will then try to reach out to sales-qualified leads; this process of initiating conversations with qualified leads is called cold-calling. The digital part of your sales strategy can complement your cold-calling and outbound email efforts. SDRs are also starting to use social media platforms as lead sources, where they can target potential leads with innovative content as well as paid advertising.

SDRs will work with **business development managers** (**BDMs**) to develop business relationships or strategic partnerships to complete the most important part of the sales process – making the deal. This will often require multiple meetings and product demos to external stakeholders.

Optimizing sales performance

Making a sales strategy is the optimization phase. Here, sales managers and representatives look at the results of their work and develop ways to improve them. It consists of team structure, time management, and sales skills.

Is your sales team set up and organized to do its best work? A typical structure should have SDRs, **account executives** (**AEs**), and Customer Success, who make sure the customer is happy after they sign on the dotted line. If this is part of your go-to-market sales strategy, you may need managers for each group of reps, and you'll want to set up separate teams to handle prospects and sales leads.

It's important to remember that both sales leaders and reps need to be able to handle their time well. The best sales teams do better when they share winning strategies and information. Learning and development can be managed by someone else, but it's best to do it yourself.

Establish data-driven tracking

This is the part of a sales strategy that looks at data. Sales managers should give their reps goals and KPIs to work toward. For the best results, individual sales rep and team goals should be aligned with larger company goals.

Lead generation efforts are measured by the number of calls made and emails sent, the number of meetings booked, and the number of sessions attended. Overall sales success measures include product demos, the **monthly recurring revenue** (**MRR**), and pipeline metrics such as average deal value, demo attendance, and the closed-won deals ratio. You will learn about these metrics in depth in *Chapter 12, Business Metrics*.

The most important KPIs for every sales team are the amount of money made from inbound and outbound conversions and the total amount earned.

Implementing sales tooling

Lastly, you'll need to choose which of the thousands of sales technology solutions you'll use to power your sales strategy. You will need three types of software: **a customer relationship management (CRM)** tool, a sales intelligence tool, and an email tool.

Your CRM is the most critical part of your tech stack. It keeps track of every time you talk to a prospect or customer, so salespeople and managers always have all the information they need. Salesforce is the most popular CRM on the market.

Sales intelligence or B2B prospecting tools help find customers who might be interested in buying. It also lets you find the information you need to get in touch with them, such as email addresses and phone numbers you can dial directly.

Sales automation software is used to schedule emails, keep track of responses, and track engagement.

Those in charge of sales should thoroughly audit their tech stack. Always try to get the most out of the technology you already have. At any given time, thousands of tools are on the market, so it's essential to do your research before picking out the right tool. In the start-up phase, the person in charge of sales will sign up for the sales automation tool and, if needed, set up training for the rest of the team. However, as the team grows, you will have to look into newer tools in an evolving landscape to make sure your team is equipped with the tools that make them productive.

As you set up a sales process with the right tools, tactics, and team, as well as KPIs that track your progress regularly, the patterns of success and failure will help you change your sales practices to build a winning sales strategy over time. In addition to building a sales strategy, a great way to reduce CAC is to build products that sell themselves. You will learn more about this in the next section, *Product-led growth*.

Now that you understand the mechanisms you can utilize for growth in terms of strategy, operations, and measurement, in the following sections you will learn about product-led growth, community-driven growth, and no-code and low-code tooling that is widely used for APIs and other SaaS products.

Product-led growth

People talk a lot about "product-led growth," which works well with SaaS products because they don't have to pay much to get new customers. Product-led growth is a strategy that involves making products in a way that makes it cheap to acquire, grow, convert, and keep customers. Customers can make a buying decision with the least amount of sales effort if you design the product to be easy to use and approachable for them, as well as allowing wider audiences to discover and try your product with the least amount of friction.

With a product-led approach, the goal is to grow a business by getting more people to use and interact with a product instead of getting new customers.

In a traditional sales-led approach, the sales and marketing teams capture leads into the sales funnel and guide them through it, via educating your users through content, personalized email campaigns, demos, newsletters, exclusive offers, and countless other strategies.

With a product-led approach, you get users into the funnel by getting them to use your software. This will skip the awareness and consideration stages and move customers closer to conversion right from the get-go.

Moving customers through the funnel with a product-led approach saves money and resources on marketing and sales-related activities.

Also, most people would prefer to try a product for themselves than listen to a salesperson's pitch about it. In addition to that, getting new clients and employees up and running takes time and money. It's not the best way to teach your employees if you only do one demo call and then use the recording to teach everyone else.

Self-service onboarding solves this problem, making it one of the most critical parts of a product-led growth strategy. Self-serve onboarding is a way for people to sign up for and pay for a product or service without talking to someone from the company.

Usually, this kind of system is made simple and easy to use, so users can get started quickly and without help.

A sound self-service onboarding system has the following parts:

- A way for users to sign up on their own
- Interactive built-in tutorials
- A help center with many articles on how to do things
- The ability to schedule a custom training session with your team if needed
- A 24/7 support channel
- A way for users to change their plan or get access to add-ons
- A way for users to stop paying and close their account

Your tool does the majority of the work for you with little assistance. However, you should still be able to give your users personalized demos.

A growth strategy based on products is better than one based on sales in several ways.

First of all, it lets a company focus on making a great product instead of trying to sell it. This can be good for the company in the long run because customers are more likely to keep using a product they like if they are happy with it.

Second, a product-led strategy can help a company build a strong brand, reputation, and word-of-mouth publicity because customers are more likely to recommend products they like.

Lastly, a strategy that focuses on a product can be more efficient and cost-effective than one that focuses on sales because it requires less investment in marketing and sales efforts.

However, the product-led model might not work for every business. The biggest problem with a product-led growth strategy is that you will have to give users a free or *freemium* version of your tool that lets them try it out for free, without any help from your team.

When you offer a freemium product, you have to accept that a majority of the people who use it will never buy anything. They will either work on that one quick project or use as much of your tool's free features as they can. This is a big drain on your company's resources because it takes a lot of infrastructure and maintenance costs to keep a large number of people using it.

But if it works out, you'll have software that sells and markets itself because your users will tell their friends and coworkers about it. This will bring in more and more potential customers in a snowball effect while keeping marketing and sales costs low.

An extension of product-led growth is community-driven growth, which is particularly effective for APIs and other developer products. You will learn about community-driven growth in the following section.

Community-driven growth

APIs have targeted developers, and the developer community is very engaged and thriving. With the increased awareness of jobs in tech, there is a large population of people who are interested in becoming developers, and the developer community is eager to help. Developer relations and developer advocacy is a function in all companies that build APIs. The entire focus of these teams is to increase education on the topic of APIs by creating a wealth of content across videos, blogs, social media, and so on.

When customers are trying to integrate with a set of APIs, they can usually turn to content that is made available to aid them, but they can also use forums such as Stack Overflow to get answers. Increasingly, API teams are beginning to follow these public forums to gain insights into customer pain points.

A thriving community also results in valuable user-generated content that is great for the customer base. A great example of community-driven growth is when certain YouTube creators develop tutorials on your APIs that are more successful than the ones created by your own team. When the community around a product grows, the product starts to grow in popularity organically. Due to the organic reach and free marketing that community-driven growth offers, it is the best type of growth and one that makes your APIs the default choice for the use cases you serve.

Low-code and no-code integrations

All the tactics for growth that we saw so far in this chapter try to reach more people, increase awareness, and get more customers. However, once a customer is interested, they have to go through the learning curve to understand how the APIs that you offer are designed for successful integration. This would require development time and effort on the customer's side, which can be time-consuming. This can also involve setting up additional infrastructure or tooling that can see customers take weeks, and sometimes months, to complete their integration.

Fortunately, you can build tooling to reduce friction for customers to integrate quickly so that they can start using APIs as quickly as possible.

No-code/low-code is a popular term for tools that require little to no coding. This allows users to use a GUI to make configurations that generate code for them. A good example of this is PayPal's integration builder where you can check a few boxes to make selections, and the tool generates code that you can copy and paste into your website for a successful PayPal integration.

Postman collections allow users to explore APIs and make modifications to API requests to validate results, without a single line of code. Users can use Postman collections to explore APIs and make changes to API requests to test results, without having to write a single line of code. Users can then copy code generated in the language of their choice and paste it into their website to complete the integration.

Twilio recognized that enterprise customers were often spending more time establishing relationships with Azure or AWS when building Twilio applications than they were on developing code. To reduce the dependency on an external serverless offering, Twilio started to offer its own set of serverless tools to enable customers to integrate with Twilio faster.

Summary

For a business to be successful, growth and adoption of its products are crucial. You can build a great product, but if nobody uses it, all your effort is wasted. To ensure that you are reaching the audience that will appreciate your product, you can now identify the target audience, size the market using a TAM calculation, and build a marketing and sales strategy for your product.

You can set KPIs and metrics for your sales and marketing teams that complement your pricing strategy and target growth to build a data-driven approach for the success of your product. You will be able to design frictionless onboarding experiences to unlock product-led and community-driven growth, thereby reducing CAC and creating a loyal customer base. You can also explore innovative no-code and low-code tools that will simplify complex products, thereby widening your audience and allowing customers to use your products more efficiently.

In the next chapter, we'll look more closely at support models for APIs and how to find and fix points of friction that customers run into during the onboarding process.

6
Support Models for API Products

Product managers are frequently eager to spend time with consumers because they act as the "voice of the customer" for the product development team. Since product managers have a number of responsibilities, they cannot talk to customers as much as they might like. However, there are teams with sole responsibility for communicating with customers. These teams are known as customer support or customer success teams.

When customer support teams are involved in the product development process from the beginning, it allows for more significant insights into customer sentiment and experience, which is an invaluable feedback loop for effective product development.

This chapter explores the connection between the API producer life cycle and the API consumer life cycle and, most importantly, customer feedback loops. The topics we will look at include the following:

1. The producer and consumer life cycle
2. Designing customer feedback loops
3. Customer feedback at scale
4. Setting customer expectations of support and **Service-Level Agreements (SLAs)**
5. Scale-based support models
6. Support metrics

The producer and consumer life cycle

As a producer of an API, it is often easy to forget that a development cycle on the customer's side is required to integrate with the APIs you produce. Unlike a mobile app or a website, the user experience of an API involves a developer or a team of developers being able to discover, evaluate, integrate, and scale their usage of the API. Depending on how easy or difficult your API is to use, and whether you provide tools to minimize the effort needed, this can take anywhere from a few minutes to a few months.

Like every product, API products need a healthy flow of customer feedback to keep the customer's experience as the top priority for product development so that the APIs can be improved over time and made better for customers. To achieve a continuous flow of feedback in API development, teams create customer feedback loops via support teams that enable the customer feedback and pain points to be recorded and reported back to the product development team for insights into the state of the APIs.

The following diagram shows the API producer on the left, representing a team that develops APIs. On the right, we see API consumers, representing customers developing applications using these APIs.

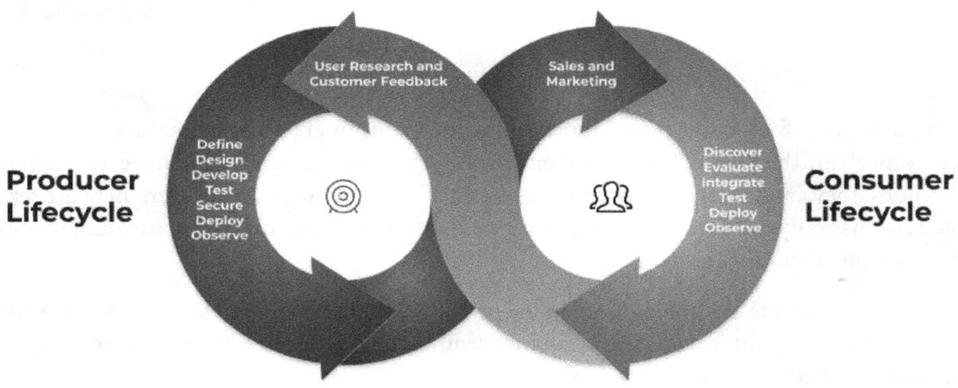

Figure 6.1 – Producer and consumer life cycle

In early phases, APIs have a limited set of customers, and it is easy to engage with the customers to get their feedback. Still, as product adoption grows, the support capacity evolves.

Existing customers could be customers who already use our APIs or are in the process of integrating with the APIs. For existing users, customer support is the only way to report issues and get help. Data generated from support generates a wealth of information invaluable for product managers in understanding customer pain points. Customer pain points get addressed with new features released to customers via sales and marketing channels for adoption. This cycle ensures customer feedback is cycled into the product development process and prioritizes future features that will truly meet customer needs.

A healthy loop consists of a producer or API development team producing APIs that consumers use and consumers are able to channel feedback into the producer life cycle, accomplished by various teams on both sides of the life cycle. On the producer side, the product managers work with user research and support teams to gather insights into customer needs and distill them into requirements for the product development teams that go through the flow of design, development, testing, security reviews, and deployment to be ultimately released to customers.

Sales and marketing teams work to raise awareness and adoption of new products and features. As customers use these products and reach out to customer support for help, the customer support team's data and metrics generate information that is fed back to product managers as gaps in the product that should be addressed and prioritized.

Now that we understand the importance of a constant flow of customer feedback, we will look at how to design customer feedback loops in terms of people and processes to scale with the product.

Designing customer feedback loops

An effective customer feedback loop is probably the most efficient way to enhance the quality of your product or service in response to the level of satisfaction experienced by your consumers. The feedback loop enables you to continuously collect, learn from, and implement the suggestions provided by your users to improve your offering.

Customer experience has become a driving factor in decision-making for customers in many cases, even more prominent than price. Hence, a good customer experience can be a crucial differentiator for a product and a competitive advantage. Customer feedback loops are designed with the following goals in mind:

1. Ensure that existing customers are engaged with the product, and their needs are being actively addressed to ensure their retention
2. Build customer trust and loyalty so they feel comfortable building with our APIs for their long-term goals
3. Reduce customer complaints by effectively addressing problems and issues
4. Analyze data from customer feedback to understand the critical issues customers intent on working with the APIs have and find out features that customers find exciting
5. Reduce the churn in the customer base

APIs' customers can be independent developers or developers who are part of a larger organization. APIs are B2B products; for such products, the support models must be developed as a combination of usage and tier-based.

For internal APIs, the developers within the organization will have access to the API development team directly to get support. For partner or public APIs, however, external-facing support channels are the only way to get help. These channels need to be designed to understand how to prioritize these requests and usually do that based on customer tier.

API companies offer paid support services that allow large customers to be able to get priority support whenever developers from those organizations reach out to your customer support teams. Customer support teams can then identify incoming requests as coming from the paid support tier and prioritize such requests accordingly.

Depending on the scale of customers, some companies have dedicated account management teams and solution architects who have a long-standing relationship with these customers and a deep understanding of their specific use cases. The support team will be able to identify these members and connect them to customers whenever those customers are in need, to provide the highest level of support.

Account management and solution architecture teams are a big part of serving enterprise customers. Account managers manage the relationship with customers on an ongoing basis and get the right technical teams involved to help customers build complex and large-scale integrations. Such contracts are handled by sales teams and can take months to establish. Large contracts and customers are essential to the growth of API products as this will entail the adoption of the APIs at scale, consequently increasing the scale of usage and revenue.

Now that we understand the goals of designing customer feedback loops and the complexity of providing support as multiple teams get involved, we will look at some support tools that help accomplish these goals in a large enterprise setting. In the following section, we will dive into what customer feedback looks like when there is a large number of users and product managers can't engage with all of them directly.

Customer feedback at scale

Early-stage products or start-ups can have a small team supporting customers. But as APIs start to have hundreds and thousands of users, it is no longer possible to use simple channels for managing support requests. For this reason, many specialized tools have been developed to build support operations.

Some of the most prominent tools used to build support systems are Zendesk, Freshdesk, Intercom, Salesforce, and HubSpot. These tools allow customers to get support using channels such as email, chat, text, WhatsApp, and so on while providing tooling for the support teams to receive, prioritize, label, and respond to incoming requests on time. These tools are highly customizable and can manage large support teams distributed across various functions and geographies.

Using an enterprise-level support tool ensures that customer requests are being recorded and addressed promptly. The data is available to be analyzed to establish metrics to measure support quality and efficiency.

There are many specialized features in support tools that enable support staff to quickly look up customer information when a customer reaches out and provide personalized support. This information could be about the support tier of the customer, their usage, the length of their relationship with the company, and their recent interactions with the support team. Such processes build a sense of continuity in the customer's experience interacting with the support team, and with reduced friction in getting help, they can get their issues fixed more quickly.

At the enterprise level, customer support involves hundreds of employees and tools to support this level of operations. These teams are established to provide a high level of support to customers. In the following section, we will learn about SLAs, which are the foundation for establishing support models in any organization.

Setting customer expectations of support and SLAs

In the previous section, we look at how large volumes of customer feedback are handled using tools to operate a support model. It is also vital that we communicate our support offering to customers so that they have a clear understanding of what level of support they can expect. This is where our concepts of API maturity from *Chapter 3* are instrumental. For experimental APIs, we can control the exposure to a limited set of customers because it is natural that the product and documentation are not thorough at this point and there would be more support needed for customers to use such APIs.

It is standard practice for API product documentation to have a page dedicated to maturity. This maps each maturity level to the level of support provided. This allows customers to have an understanding of what to expect when they reach out to support for help. This mapping also involves expectations of how frequently APIs may change or whether any APIs are on the path to deprecation. Customers look at these pages to plan development and manage dependencies on their end to make sure their integration is stable and their customers are not impacted by changes to the underlying APIs.

For large customers, there are complex contracts in place that ensure a high level of support, along with a dedicated technical solution architecture team and technical support. Occasionally, product managers are also involved in advising large customers when high-impact features are released to make sure those customers have appropriate SLAs as part of their contract with the API producer. In the following section, we will dive deeper into the operations details of support requests and understand how customer support requests flow through support queues, get escalated when needed, and flow back into product priorities to enable a continuous customer feedback loop across the product development life cycle.

Scale-based support models

For internal APIs, internal developers have the luxury of reaching out directly to the product and development teams supporting those APIs, but for partner and public APIs, the support channels are more formal. In this section, we will look at support flows for public and partner APIs.

In *Chapter 3*, we learned how API maturity is presented to and perceived by customers. API maturity is closely tied to customers' expectations of the quality of the product and the maturity of any associated features or services. In the following diagram, we can see the operating model for a maturity-based support flow for large-scale API products.

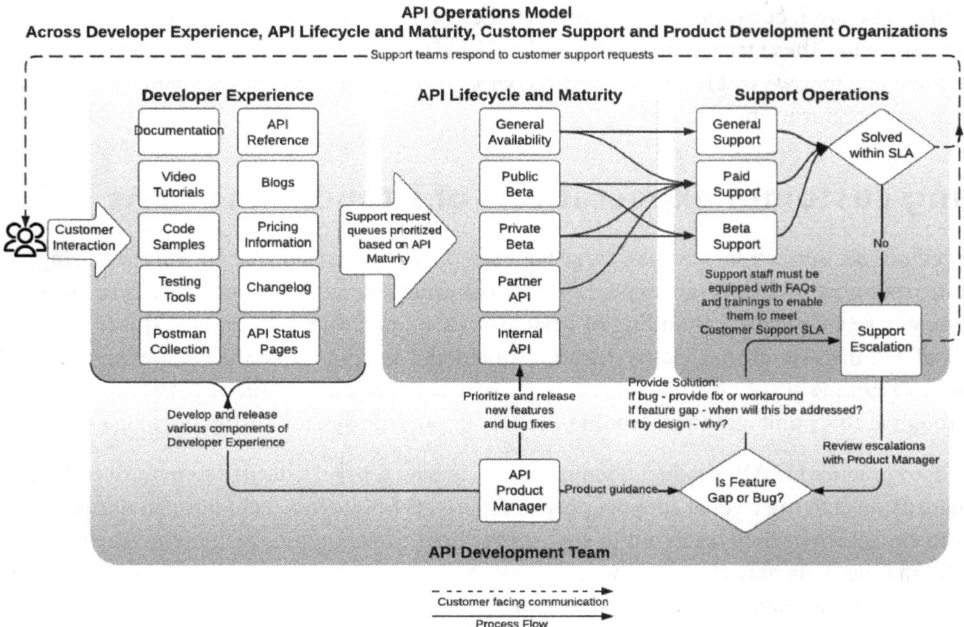

Figure 6.2 – Maturity-based support flow for large-scale API products

To understand the support flows, we must understand the touch points for the customer. Customers interact with the web pages and support channels listed on the website that are part of the developer experience. The developer experience includes components such as API reference pages that list all the endpoints of the API along with parameters and code snippets in depth. Documentation for an API can include use cases and sample code. Usually, touch points for developers expand beyond API references and documentation to include blogs, videos, changelogs, pricing pages, testing tools, API status pages, and Postman Collections.

The first thing to keep in mind is to make sure there is a way for customers to find a support channel on all these touch points so that if and when they are in need of support, they have a way to reach out and get the help they need. As customers reach out for support, all customer requests get queued up in the support tooling and are then prioritized and assigned for triage and response. These are known as **support queues** and are usually designed to match incoming customer requests with the appropriate support staff who are equipped to respond to customer requests effectively so that customers get quick and correct responses.

Every component of the developer experience should map clearly to the API maturity model. Most public-facing documentation is expected to belong in the generally available category on the maturity scale, which means the majority of tickets are expected to be created for generally available APIs. The status of general availability for an API means that support is provided at a high level of SLA.

For customers that have additional paid support services, the ticketing system must recognize the user ID and route the support request to a prioritized queue for paid support. Product managers across the organization ensure that in addition to the public-facing documentation, there is also internal training available to support staff so that they are equipped to answer incoming customer requests.

Beta products are not mature enough for support staff to answer questions directly; support requests for beta products have to be handled through escalation to product teams. Customers using beta products expect a slow response and the beta flag is designed to ensure that this expectation is set. Requests pertaining to beta products flow through a beta queue, which ensures that the support team gets a chance to escalate and review requests without impacting the SLA for general availability support queues.

As APIs are developed in an iterative manner through the course of the product life cycle, as support requests come in, not all of them can be addressed immediately by support staff. These requests would have to be reported to product managers of the respective APIs to get their input on how best to respond to the customer. Product managers review support escalations on a weekly basis to understand the nature of questions users ask and whether these can be addressed with new features or bug fixes, or whether they are by design. The solution path is then documented for future requests of the same nature for a quick turnaround and feature requests and bugs are added to the product backlog. This way, the product evolves with customer feedback and customer support improves as the knowledge base is improved over time.

Product managers would also take learnings from customer support requests to improve developer experience components over time, which could include adding sections to the documentation and creating videos or blogs. As new features are launched to address product gaps, customers who reported or requested those features can also be informed, using the information on support requests, so that they can be made aware of the features that they were excited about. This cycle of capturing customer requests, funneling through support, and reporting on them ensures a healthy feedback loop that keeps products evolving even at the scale of millions of users.

Support metrics

There are a few different metrics that can be utilized to measure whether the support model we designed is working as desired. The goals of establishing a support model include the following:

- Customer requests are received, triaged, and responded to
- Customers get a satisfactory response to their requests in a timely manner

You can use a variety of metrics to measure various aspects of these two goals. In the following section, we will review the most commonly used support metrics across the industry.

Ticket volume

The most high-level metric that is measured is the raw volume of support requests coming in. The volume of support tickets will be directly proportionate to the number of customers the product has. This volume will determine the most fundamental need for staffing support teams. As organizations grow, they might have a vast number of products with thousands or millions of users. You can imagine the need for staff to handle the requests coming in growing exponentially.

Monitoring the number of tickets is important for planning how to divide up the work among support agents, figuring out when you might need more agents on hand, and finding gaps in coverage. Total conversations is a metric of this type and includes all interactions with customers, whether they happen through an official support ticket, a Facebook message, or some other social channel.

Track how the number of tickets changes over time to find out what times of the day and days of the week your team gets the most requests for help. With this information, you can plan shifts and divide work more efficiently. This will not only make sure that your support team isn't too busy, but it will also help your team respond faster, so customers don't have to wait as long and can get their problems fixed more quickly.

Customer satisfaction (CSAT) score

Customer satisfaction (**CSAT**) is a way to measure how happy your customers are. It's based on a survey that customers usually fill out after talking to your support team. Even though the survey's questions can be different, it always asks customers to rate their experience on a scale. This helps you measure customer satisfaction, which is often a qualitative metric.

CSAT is an important metric for customer service. Customer support teams' main goal is to help clients and make sure they are happy. CSAT measures how well these efforts work and gives direct feedback from the customer.

The following equation shows how to calculate the CSAT score:

$$[\frac{Number\ of\ positive\ responses}{Number\ of\ total\ responses}] \times 100 = CSAT$$

Tracking CSAT scores over time is a good way to measure the consistency of the support provided to customers. After any major releases when major product changes are made, monitoring CSAT scores can be very insightful in identifying whether the release has the desired impact.

First response time

No one wants to wait a long time to hear back about a problem they're having, so if you can keep your response times short, your customers will be happier. First response time is the amount of time between when a customer submits a support ticket and when the first response is given by a customer service representative. The length of time a customer has to wait before getting help is a good indicator of how well your support team works.

The following equation shows how to calculate "first response time":

$$[\frac{Sum\ of\ First\ response\ times}{Number\ of\ tickets}] = Average\ First\ Response\ Time$$

You can keep track of how long it takes each customer service representative to answer the first time, as well as the average for the whole team. If the average is too high, you can see who might need more help or training by keeping track of how long each agent takes. Zendesk says that a good first response time for email is 12 hours; for social media, it's 2 hours or less, and for live chat, it's an hour or less.

Companies usually measure this time in business days, so weekends and off-hours won't slow down your average response time.

Average resolution time

Average resolution time (**ART**) is the average amount of time it takes your team to handle a support ticket from the time it's opened until it's closed and the problem is fixed. Long resolution times mean that customers have to wait longer, which can affect how satisfied they are and, in the end, how long they remain customers.

There are industry benchmarks for how long it takes to solve certain kinds of problems. API products might not match the industry standards due to the unique nature of the products. You need to establish standards based on your own company's benchmarks and customer expectations.

If your weekly ART is high, it could mean that your current support methods aren't working well or that your agents have too much work to do. Talking to your support team to find out where they are having trouble can give valuable insight into areas of improvement in the product.

Segmenting the average time it takes for each agent on the team to solve a problem will let you know whether any team members are having trouble.

Here is how to calculate the average time to solve a problem:

$$[\frac{Total\ time\ to\ solve\ all\ tickets}{number\ of\ tickets\ solved}] = Average\ Time\ to\ Solve\ Tickets$$

Check the average time to resolve issues to see whether anything has changed. A steady rise over time could mean that your support team needs more help or training to deal with customers' new, more complicated problems.

Average handle time (AHT) is another customer service metric you might track. This shows how long an agent spends on average on a single call or chat with a customer. AHT doesn't include the time a customer waits for an agent to pick up their ticket and make the first contact, which is part of the ART.

In addition to the ART, when analyzing support metrics for an API, it's a good idea to also look at percentiles. Percentiles, such as the 70th and 90th percentile, can provide a more complete picture of the performance of the support team and can help to identify potential issues that may be hidden by averages and medians.

The average resolution time provides a general idea of how long it takes for the support team to resolve issues, but it may not take into account outliers or extreme cases. The median is a better indicator of the typical resolution time, but it doesn't give you an idea of how long it takes for the longest issues to be resolved.

Percentiles, such as the 70th and 90th percentile, can help to identify the longest resolution times and can help to identify early signs of problems or areas for improvement. For example, if the 90th percentile resolution time is significantly longer than the average or median, it may indicate that there are some issues that are taking a long time to resolve and that the support team needs to focus on these issues to improve the overall performance.

It's important to remember that these metrics should be used in combination with others, such as the number of open tickets, the number of tickets resolved per day, customer satisfaction, and customer retention.

Ticket volume to active user volume

In addition to the metrics we saw previously, you can create derived metrics that can shed light on other aspects of customer experience. The ticket volume on its own can help you evaluate staffing decisions, but if you see the ticket value as a proportion of monthly active users, you can evaluate the growth of the product against the cost of support.

You can also establish the baseline of how many customers are able to use the product without the help of support on a regular basis and see whether that ratio is relatively consistent over time. Spikes or dips in this ratio would be signals for product changes or seasonality that must be kept in mind to better understand and evaluate customer support.

Segmentation of tickets

For large organizations, understanding the ticket volume across the different dimensions is critical for making various operations decisions. For example, if most of the support tickets are in the paid support category, that is acceptable customer behavior since the customers are willing to pay for that support service. But at the same time, it is important to understand how the requests coming in from this category differ from other categories and see whether that aligns with the product's expected behavior and strategy.

Segmenting support tickets based on severity and priority can play a crucial role in understanding and analyzing support ticket metrics.

Severity is a measure of the impact of an issue on the customer or the business; it can be used to determine the urgency of the issue and the resources needed to resolve it. For example, a critical issue that affects the entire system would be considered high severity, while a minor issue that only affects a small number of users would be considered low severity.

Priority is a measure of the importance of the issue and the order in which it should be resolved. For example, an issue that affects a key process would be considered high priority, while an issue that has a low impact on the business would be considered low priority.

Segmenting tickets based on severity and priority can help to identify patterns and trends in the types of issues that are being reported and can help to identify areas where the support team can improve their performance. For example, if the support team is consistently resolving high-severity tickets quickly, but low-severity tickets are taking much longer to resolve, they may need to rethink their approach to prioritizing tickets and allocating resources.

Additionally, segmenting tickets based on severity and priority can help to identify which issues are the most important to the business and customers, and can help to prioritize the team's efforts and resources. This will help to ensure that the most critical issues are resolved quickly and that the support team is providing the best possible service to the customers.

Geographical segmentation is also a common type of segmentation for large organizations. In many cases, there might be a separate support team for each geography served by the customer support organization, to better match customer time zones and language preferences. In the case of APIs, because API documentation tends to be available in a limited number of languages, it is also common to see ticket volumes vary between geographies where supported languages are more widely spoken and where they are not.

In the case of companies such as PayPal whose APIs are integrated by developers across the world, it is not possible to provide support and documentation in every possible language, but these tickets are still tracked and responded to in the best possible manner to ensure that all customers get the optimal level of support and can be successful using PayPal APIs.

The customer feedback loop is the final and most important handoff for the product manager of an API product as they enable support teams to support customers through their integration journey. Support tickets serve as a gold mine of information to understand real-world issues that customers are facing and, if fixed, can bring immediate business impact.

Summary

In this chapter, we reviewed the importance of the customer feedback loop in delivering a great customer experience and fueling the prioritization of the right products and features for the continuous improvement of APIs. We learned about designing scalable support channels for API products and the various teams that you partner with in the process. We also learned to tie the maturity of APIs to the level of support that a customer can expect and also how this expectation must be communicated to customers.

Now that you understand the value support data can bring to the product development process and the metrics that can be used to measure the supportability of a product, you can use these channels to better understand the customer journey and evaluate customer experience metrics.

In *Part 1*, we looked at all the aspects of being an API product manager and building API products that grow and scale. In the following chapters, we will start to look at the developer journey to learn about the customer's perspective of using APIs and how they go from discovering a new API to evaluating them for usage, integrating them into their applications, and eventually scaling their usage of APIs. Understanding each step of the developer journey will help you build empathy for the developer persona and also understand the various touch points that customers have in using APIs. We will then use this understanding to establish a strategy for delivering an excellent developer experience.

Part 2: Understanding the Developer

User research, getting to know customers, and mapping a user journey are all essential parts of the product development process because they help you make products that meet customers' needs. Doing user research in a methodical way, getting to know your customers, and making a map of the user's journey are all important steps in building successful products.

User research lets you learn about the needs and pain points of your target customers. This helps you make decisions about your product and makes sure that the final product meets the needs of the people who will be using it. Customer empathy is important because it allows you to understand customers' perspectives and how they interact with the product. This understanding can lead to more effective design decisions and a better user experience.

Mapping the user journey, or the process a customer goes through when interacting with a product, helps you identify potential pain points and opportunities for improvement. By understanding the entire customer experience, you can create more intuitive and user-friendly products.

One example of how user research helped build a successful API is the development of the Stripe API for online payments. Stripe's founders realized that, while there were many payment solutions available, they were difficult to use and integrate into websites and mobile apps. They conducted user research and found that developers were looking for a simple and easy-to-use API for online payments.

Based on this research, Stripe set out to create an API that was easy to integrate and provided clear documentation and robust support. They also made sure that the API was flexible and could be used for a wide range of payment scenarios, from simple one-time payments to complex recurring payments and subscriptions.

The Stripe API was well received by developers, who appreciated its ease of use and flexibility. It quickly became one of the most popular online payment APIs and is used by a wide range of businesses and start-ups to process payments on their websites and mobile apps. Stripe's API was a successful product because it was built with the user in mind through research and understanding the needs and pain points of their target customers.

Walking in customers' shoes

Mapping the user journey involves identifying the key steps that a customer goes through when interacting with a product, from initial awareness to post-purchase support. This includes understanding the customer's needs, goals, and pain points at each stage of the journey, as well as identifying any potential roadblocks or friction points that could impede the customer's progress. By understanding the entire customer experience, you can create products that are more intuitive and user-friendly and that better meet the needs of your customers.

Walking in the customer's shoes is an approach that encourages you to put yourself in the customer's position and see a product from their perspective. This can be done through user research, such as interviews or surveys, or by observing customers using the product. By gaining a deep understanding of the customer's needs, goals, and pain points, you can create products that are more tailored to their customers' needs.

Mapping the user journey and *walking in the customer's shoes* are both important approaches that help you create products that truly meet the needs of your customers. It is also important to note that these practices can also be applied to other areas of the business, such as customer service and marketing, to improve the overall customer experience.

Walking in the customer's shoes when it comes to APIs refers to understanding the perspective and needs of the API users, which are typically developers. This approach encourages you to put yourself in the developers' position and see the API from their perspective.

When *walking in the customer's shoes*, you should consider the following aspects:

- **Usability**: How easy is it for developers to understand, use, and integrate the API into their own systems?
- **Flexibility**: Does the API provide enough flexibility to handle different use cases and integration scenarios?
- **Robustness**: How reliable and scalable is the API?
- **Support**: Is there adequate documentation, tutorials, and support available to help developers with any issues they may encounter?

By understanding the needs, pain points, and perspectives of the API users, you can create APIs that are more user-friendly, flexible, robust, and better meet the needs of developers.

It's worth mentioning that it is also important to continuously gather feedback from API users and make adjustments to the API as needed. This could be done through conducting surveys, monitoring API usage, or direct communication with API users.

Meet the customers where they are

Customer research is a vital step in building experiences that truly serve customers. It helps you to understand your target customers and create products and services that meet their needs and preferences. Additionally, by gathering feedback and continuously iterating, you can ensure that the experiences they build are continuously meeting customer needs and preferences.

Customer research can take many forms, such as user interviews, surveys, focus groups, and usability testing. This research can be used to gather information on customer needs, pain points, and preferences, as well as gather feedback on existing products or services.

In *Part 2*, you will learn in depth about the techniques and best practices for conducting user research for API products, developing an understanding of your customers, and building API experience components that delight your customers. The following chapters will be covered in this part:

- *Chapter 7, Walking in the Customers' Shoes*
- *Chapter 8, Customer Expectations and Goals*
- *Chapter 9, Components of the API Experience*

7

Walking in the Customer's Shoes

The rising interest in API development brings up the need to think about the API-building process in a more methodical and strategic way. The first thing to look into when you consider building any product is to identify the market opportunity and conduct user research to develop a deep understanding of your target customer. Customers and their experiences are the focus of an outside-in product development approach. In order to succeed, this strategy is based on the idea that offering real value to customers is the only way to go.

Understanding your users begins with understanding their journey. Drawing a user journey map is a formal way of understanding how users arrive at a goal and the steps they take to get there. Developers are making more and more decisions about which technologies their organizations should use. As you build APIs that are mostly used by developers, it's important to map out the path your developers take from the time they first find your APIs to the time they help them grow. Once you map out their journey, you can start to dive into and analyze each step to more deeply understand the points of engagement as well as points of friction. If you know a lot about the developer's journey, you can find ways to make their overall experience better.

In this chapter, you will learn about the following topics:

- Prioritizing user research
- Establishing user personas
- Mapping the developer's journey
- Determining customer touch points
- Identifying points of friction and conversion

By the end of the chapter, you will have learned how to put customer needs at the center of your product strategy by developing a user research methodology that is useful for product prioritization and effective communication of customer needs across the development team and stakeholders.

Prioritizing user research

In *Chapter 5, Growth for API Products*, you learned how to identify the target audience and how to segment your target market. Once you've chosen a market based on a high-level strategy, the next step is to dig deeper into that market and start doing user research to learn more about all the user personas in that market.

User research is a close partner team for all **Product Managers** (**PMs**). When building API products, you can start by first learning who your customers are. You can do this by establishing customer personas and mapping the customer journey for each of these personas. A developer's journey is a map of sequential steps across all your documentation, tools, and marketing channels, such as blogs, videos, and others, that your customers use to learn about your product offering, assess whether it is a good fit for them, develop and integrate with your APIs, and use the product actively.

The primary users of the APIs you build are going to be developers. However, developers may not always be the only users. There are several other team members who are not developers but who are also part of the decision-making when it comes to choosing a set of APIs for the purpose of their project. This could include the product manager, security professional, software architect, and so on. In a **business-to-business** (**B2B**) setting, different stakeholders make decisions, and each one needs a different set of information to evaluate your API offering. In the same way that there are several teams and team members involved in the development of APIs, your customers also have a number of team members who collaborate in the process of discovering, evaluating, integrating, deploying, and monitoring their integration using the APIs you produce.

It is the responsibility of UX researchers to generate trustworthy insights that can serve to direct the decisions made by product teams. Good intuition is an asset in the process of developing user-friendly and impactful products. The product development process is heavily reliant on UX research insights, and poor UX research insights can lead to poor product decisions based on facts and/or conclusions that are inadequate, inconsistent, or erroneous. On the other hand, it's up to PMs to correctly evaluate these insights and use them to put a strategy into action.

Prioritizing the research process starts with asking the right questions that user research efforts can help research and shed light on. You might have a long list of research ideas across various aspects of the product experience, such as the following:

- What are the various customer personas who use your product?
- How do users discover or become aware of your API offerings?
- How do users compare your APIs to competing API offerings on the market?
- Is your pricing model working for your customers?
- How can you drive upgrades and increase retention?
- What developer tools do your customers most commonly use?
- What do your customers' tech stacks look like?

All of these are important questions to answer in order to understand the customers and opportunities for your product offering. If you ask your users clear, simple questions, your research will be more useful and effective. User research should be a continuous part of your product development process, and this means that not only should you have user research that is in progress that is going to fuel the next iterations of your product, but also user research that validates your current understanding of the user base so that you are aware of the changing marketing dynamics and user sentiment.

Big roadmap items can sometimes overwhelm your research strategy, but if you spend too much time on long-term projects, you might miss out on ways to help your customers more immediately. On the other hand, if you work on a lot of small goals and changes, the **user experience** (**UX**) could become busy.

To balance your user research efforts against short-term and long-term goals, fill your schedule with mostly short-term goals that meet both the company's current strategy and the needs of users right now, with long-term research needs spread across longer periods of time. If you're still not sure which project needs your attention, you can run a cost of delay analysis, which looks at how much potential revenue you'll lose by waiting. This comparison of timelines and effects can help you figure out which projects could pay off the most.

Your user research strategy should enable you to go broad and understand things such as the size of the business, geography, industry, and so on, that your customers belong to. Once the broader segments have been identified, you can go deeper into each segment to understand trends such as the size of the team, skills on the team, and the tech stack being used by your customers.

When creating a user research strategy, it is important that you look at a wide range of customer segments so that you can not only identify the ideal customers for you at the moment but are also able to create a prioritized set of user personas that you might want to target in the future. In the following section, you will learn more about creating user personas that can be shared within a team to build a shared understanding of customers.

Establishing user personas

When building consumer products in a B2C environment, you can segment your audience in a handful of ways, such as demographics, age groups, location, income buckets, and so on. This lets you figure out which groups of people like or dislike your product. Similarly, in B2B and B2B2C environments, you can segment your audience into the following categories:

- **Individual Developers**: In a number of cases, your customers might be a small team of a handful of people, or even a single person, who are building applications using your APIs. A common scenario is where there is a single developer in a university or a non-profit organization. These developers make all the decisions regarding the evaluation and usage of your APIs in their organization.

- **Small-to-Medium Businesses:** Start-ups and small businesses sometimes have a small team of developers who start out by using your APIs. These teams might have a small number of people who are involved in the decision-making and development process. Small-to-medium businesses usually have a business owner who is focused on the business aspect and hires a team of developers, either in-house or externally, to do the development. These developers will only be focused on the technical evaluation of the solution.
- **Enterprise:** As you look at enterprise customers, the number of decision-makers grows significantly. The evaluation decisions that are often made by a single person in a start-up are made by a number of teams in an enterprise setting. The scale of usage is also significantly larger and you will partner with sales and account management teams to connect with this type of customer.

Any development team could have one or more developers with different levels of experience and areas of expertise. Depending on the skills available on the team, they will look at solutions and tooling options. For example, if a customer is an individual business owner, they might prefer no-code or low-code solutions because they are easier to maintain. While if the customer is an enterprise user with a number of senior developers on their team, they might have questions about scaling and security that other customers may not have.

You can also segment your audience based on industries such as finance, healthcare, and so on. Once you establish the segmentation of your customers, you must create a user research strategy in a way that allows you to speak to a variety of those segments. When you work on an existing product, you can run online surveys to allow users to give feedback on your product and also input that allows you to segment your audience. Existing users are a great way to get insights into customers who are already successful in using your product.

Customer interviews are a great way to get deeper insights into your customer base. You can reach out to existing customers and incentivize them to meet with you for a 45–60 minute conversation, or you can use user research teams to help recruit potential customers. Work with your user research team to create a user interview questionnaire. This will give you a list of questions you can ask customers that are organized and in the right order. This will let you create a dataset that can be mined with answers from your users about the different things you ask them about.

Diary studies are a great technique to get insights into the actual integration process that your customers go through in the process of integrating with your APIs. With this technique, you will recruit developers and share a specific set of instructions with them. Developers will keep a diary where they take notes as they follow the set of instructions, and at the end of the exercise, they will share their notes with you. This is usually done over a few days to give developers time to work independently, and you conduct interviews at the beginning and end of the process.

As you talk to customers through various channels, you can create a methodology to log and track all the learning in such a way that you can compare and contrast the responses from various customers and use this information to identify patterns. This clustering of information will help you develop archetypes of users that are called "user personas" since they represent the demands of a wider segment of the population. One- or two-page documents are usually all that is needed to convey a persona (like the one you can see in the following example). Examples of one- to two-page summaries include a person's behavior patterns and goals, as well as abilities, attitudes, and background information. When creating user persona templates, designers often incorporate some fictional personal details (for instance, quotes from real users) as well as context-specific details.

A set of well-researched personas helps make them more relatable and enables your team to collaborate and ideate together. The following figure shows what a common user persona template looks like for four customer personas:

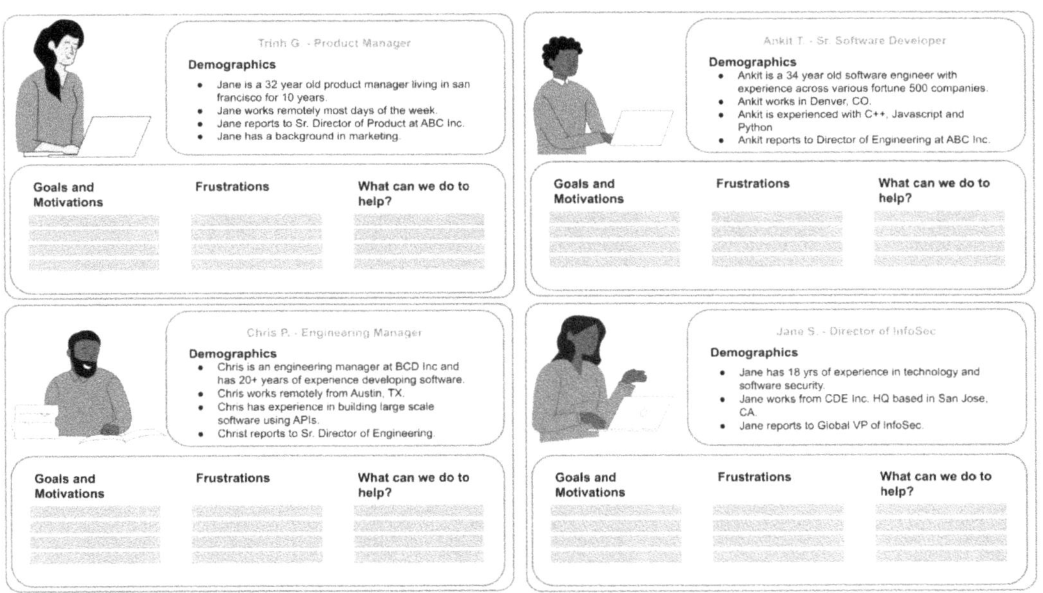

Figure 7.1 – Examples of user personas

You will create a number of personas for each segment of your audience. It is common practice to abstract personal information from user personas and use fictional names for privacy reasons. This allows personas to be long-term artifacts for a product that can be used by the development team as well as any stakeholders on an ongoing basis. When you prioritize features, you can ground your design in the person the feature is targeted toward. Support teams can also use personas to communicate the pain points they are observing customers having.

Using user personas, you can start to map out the user journey that identifies various steps the users take to discover and use your product. You will learn about mapping the developer's journey in the following section.

Mapping the developer's journey

The path a customer takes while discovering, evaluating, and ultimately using your product is known as the "user journey." A user journey map depicts the user journey in a visual format alongside the user's activities, goals, touch points, and responses.

User journey maps are very useful in understanding how your customers connect with your brand, product, or service. User journey maps vary from company to company and might vary from industry to industry. However, they always show a timeline of the user's trip that summarizes the most crucial steps.

As you apply the user journey mapping methodology for API products, you will notice that "user" and "developer" are often used interchangeably. In the following diagram, you will see how to map the user journey across discovery, evaluation, integration, testing, deployment, and observability mapped to the various questions customers have, customer touch points, the activity customers complete in each step, and how they respond to the current offerings that enable them at each step.

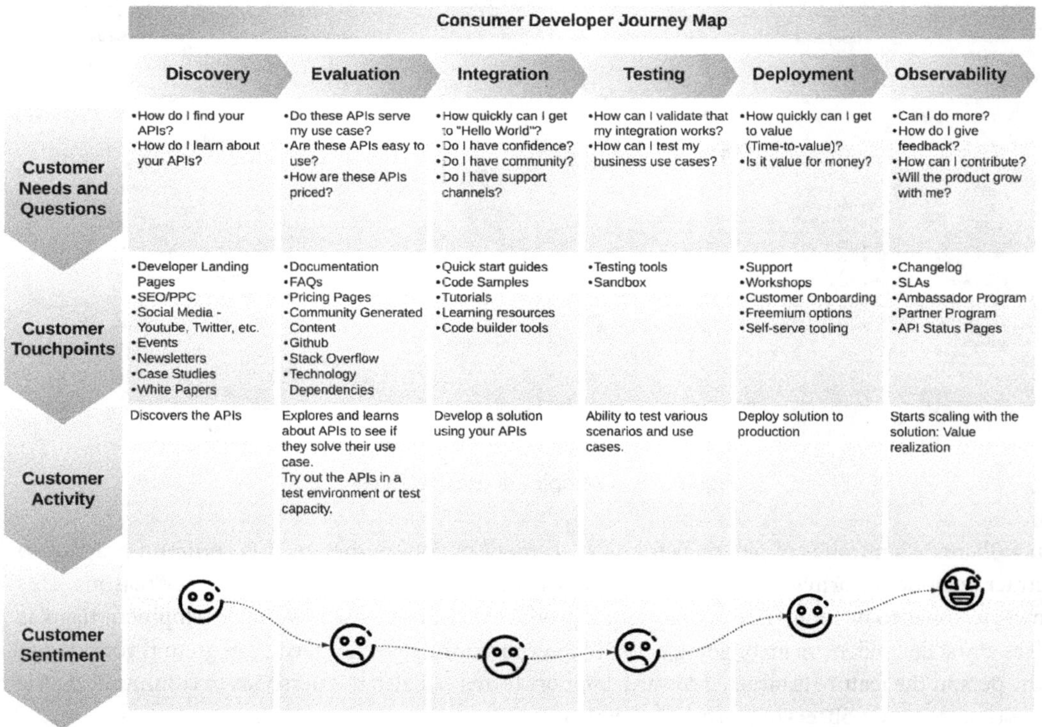

Consumer Developer Journey Map
Discovery
Evaluation
Integration
Testing
Deployment
Observability
Customer Needs and Questions
•How do I find your APIs?
•How do I learn about your APIs?
•Do these APIs serve my use case?
•Are these APIs easy to use?
•How are these APIs priced?
•How quickly can I get to "Hello World"?
•Do I have confidence?
•Do I have community?
•Do I have support channels?
•How can I validate that my integration works?
•How can I test my business use cases?
•How quickly can I get to value (Time-to-value)?
•Is it value for money?
•Can I do more?
•How do I give feedback?
•How can I contribute?
•Will the product grow with me?
Customer Touchpoints
•Developer Landing Pages
•SEO/PPC
•Social Media - Youtube, Twitter, etc.
•Events
•Newsletters
•Case Studies
•White Papers
•Documentation
•FAQs
•Pricing Pages
•Community Generated Content
•Github
•Stack Overflow
•Technology Dependencies
•Quick start guides
•Code Samples
•Tutorials
•Learning resources
•Code builder tools
•Testing tools
•Sandbox
•Support
•Workshops
•Customer Onboarding
•Freemium options
•Self-serve tooling
•Changelog
•SLAs
•Ambassador Program
•Partner Program
•API Status Pages
Customer Activity
Discovers the APIs
Explores and learns about APIs to see if they solve their use case.
Try out the APIs in a test environment or test capacity.
Develop a solution using your APIs
Ability to test various scenarios and use cases.
Deploy solution to production
Starts scaling with the solution: Value realization
Customer Sentiment

Figure 7.2 - An API customer's user journey map

The path of a developer is made up of the five steps we'll talk about in the following sections.

Discovery

The first step for a user is to discover your product. This could happen through a number of channels, such as from a user googling your APIs, coming across your APIs as part of a YouTube tutorial about a project that uses your APIs, and so on. If you already have users and want to bring in new ones, you might already have ways for people to find out about your APIs. For example, you could use newsletters, events, social media channels, and so on. The discovery phase is very important for letting people know about your API and getting new customers interested in it.

During the discovery process, you will often find different user personas interacting with all the content you publish. Product managers on your customers' side would be looking for potential solutions based on a particular problem they are looking to solve. For example, a product manager at an e-commerce company might look for payment APIs that offer the ability to accept Apple Pay so that they can enable Apple Pay as a payment option for their customers.

At this stage, the focus of users is going to be on business use cases, and this is the top of the customer funnel for your product. As you start to think about how your customers find your APIs, you can also start to think about how developers start to learn about the capabilities and if there are ways you can enable this process.

Evaluation

Once a user comes to know about your APIs, the next step is to start evaluating them to establish whether they can be used for their purpose and whether they align with the customer's goals. From the perspective of business goals, this is the most crucial step in the user journey. At this stage, users compare and contrast your product with other competing products. They look at things such as pricing, technical dependencies, and the developer community around your APIs.

In your developer documentation, you should put things such as pricing, where the business operates, and technical limitations as close to the top as possible. Putting key information that's crucial for decision-makers in an easy-to-find location will help your customers get answers quickly and without friction. Enable customers to reach out to customer support in this phase to make sure that if they have any specific questions, they can get support as quickly as possible. By making it easier for customers to make important decisions, more of them will be able to move on to the next step of integrating with your APIs. This will also give you a chance to find out what questions your users are asking when evaluating your product.

The developer relations team is an important driver in generating content that helps customers in this phase. Developer relations teams work with developers to make content such as video tutorials, blogs, and events based on common use cases. You can also reach out to the developer community through platforms where developers are already active, such as Stack Overflow, GitHub, and Postman, to make it easy for them to learn about how to use your APIs. When developers see that there is a lot of activity on these platforms for your APIs, it gives them confidence that if they run into problems when integrating with your APIs, they will be able to find help.

Developers who have previously integrated with your APIs will frequently create content to educate other developers. For example, a number of video tutorials for PayPal APIs on platforms such as Udemy and YouTube have been created by developers who are not affiliated with PayPal. These developers are helping each other learn by sharing what they've learned and what they've done. Finding ways to encourage developer-generated content can be an accelerant for your API's growth.

Integration

Once the business use case has been identified and the developers have learned about your APIs, they will start the integration process. In small companies, this might be a single developer who builds and tests the app, but in big companies, it might be a team of developers, a product manager, and other stakeholders. In this step, the developers on the customer side are working on building and testing their application, and in this process, they will be doing the documentation and tooling in depth.

During the development process, the developers on the customer team might use the resources you give them or other resources such as Stack Overflow, GitHub, YouTube, and so on. Customers expect the resources provided by official sources to be more up-to-date and accurate compared to community-generated resources, and you must make sure that is the case. If customers come across inaccurate information in the process of their development, they might not be able to integrate successfully or it might take them longer than expected, which would increase the cost of integration.

You can work with the Developer Relations team to create up-to-date content and find platforms where you can publish repositories that your customers can use as references. Having public GitHub repositories is a common way to make it easy for customers to integrate. Postman public collections and workspaces are also common ways to help customers find and use your APIs for development.

If you provide quick-start guides or copy-paste code samples to your customers, this process can be simplified for them, effectively reducing the time they spend on this step.

Testing

Testing is an important step for any customer developing applications using your APIs. At this step, the customer could also be given tools to help them do scenario-based testing. Testing an API integration is focused not only on scalability testing but also on business use cases. Test engineers will often work with business stakeholders to determine the scenarios that should be tested. API producers will often provide a variety of tools to enable testing for consumers.

When customers use Stripe APIs to process payments, they can use transactions with credit card numbers to test how well their integration works. These test credit card numbers are provided by Stripe to enable their customers to test various card providers such as Visa, MasterCard, and so on, as well as to test for various negative scenarios, such as invalid customer information or zip codes, so that customer applications can be tested for the user experience in all these scenarios.

PayPal has a different way of setting up the same testing features. They offer a sandbox environment with buyer and seller accounts where you can process transactions and simulate the production environment without using real money. This gives the testing engineers a full testing environment where they can do quality analysis on complicated business situations such as returns, refunds, and so on.

You can also enable testing for your customers using industry-standard tools such as Postman. The goal of this phase is to give customers confidence in their integration so that they can get ready to deploy their applications.

Deployment

Once customers have completed development and testing, they are ready to go live with their applications. This may or may not be something that is recorded as a data point in your system. For example, by integrating with PayPal APIs, you can develop and test your application using a sandbox, which is a replica of the production environment but doesn't move money in the real world. It is not connected to banks or financial institutions in any way. But in order to go live in production, you have to submit documents and get your account eligible to go live. In finance, this is necessary to make sure there aren't bad actors using PayPal APIs for purposes other than intended. The data generated as a result also allows PayPal to know when a customer goes live with their application. This may not be the case for APIs in every industry, but designing a user flow that allows users to be explicit about going to production is ideal.

Just like the way the development of an API takes a few sprints, the development cycle can vary from customer to customer. Development cycles are not always representative of how small or large the development team is. Also, depending on the complexity of the solution the customer is implementing, the time from their experimental API calls to scaled usage might be anywhere from a few days to a few months.

As your customers start to ramp up their usage of your APIs, you should be able to see the invocation volumes of the APIs start taking off. Once a customer has put their application into use and increased how much they use it, they can only do routine updates to their integration. But they will be monitoring their usage, errors, and billing. This leads to observability, which we will look into in the next section.

Observability

Once a customer has successfully integrated with your APIs, they are going to start monitoring their usage. Usage is made up of the number of API calls, or invocations, and the number of business transactions. For example, for a user to pay with Venmo on their Uber account, there might be several API calls being made for each transaction. Customers track the number of calls along with the number of transactions being processed. An error in a single API call can be the cause of a failed transaction.

Customers also closely monitor the error rates for the APIs they use. This helps them monitor the experience their application is delivering to their users. Not only should your APIs have effective error messaging, but this error logging must be available to the customers in a way that they can find it easily so that they can debug their applications effectively.

Customer teams will also have admins who may or may not look at technical details of errors or usage as much as they look at billing information. The developer portal you design for your APIs must address the need for customers to be able to get detailed billing easily. There are several pricing models that are used for APIs, but irrespective of the pricing model, price transparency and reporting needs will be important aspects of making your APIs usable for your customers.

As you learned in *Part 1, The API Landscape*, APIs are built in an interactive way. This means that you will be adding features and making changes to your APIs. Existing customers can look to the changelog to get information on the changes. Changelogs are an important way to keep in touch with your customers and let them know about any changes that might affect how they use your APIs. Keep in mind that any change that requires development effort on their end or could break their application must be communicated with enough notice so that they can discover the information and plan for the impact.

It is common for all APIs to have a status page that customers can check to see if the APIs are having any problems or have had problems in the past. In instances where there is an issue with your API, customer applications might be impacted. A customer would discover this issue when their customers start facing issues, and a status page will allow them to validate whether the issue is on your end or theirs.

Now that you are familiar with the steps your customers are going to take to integrate successfully with your APIs, you can start thinking about the variety of resources that you provide your customers at each of these steps to make this process easier and frictionless. The following section dives into how identifying touch points at each stage of the customer journey can help you shape the strategy for your APIs.

Determining customer touch points

The stages of the customer journey can also be viewed in the form of a customer funnel, with customers entering the funnel at discovery and moving step by step toward evaluation, integration, deployment, and ultimately, observability. From a business standpoint, you want to enable customers to move from one stage to the next as smoothly as possible.

There are several touch points throughout the customer journey, and we touched upon a few of them in the previous section. When you design your API experience, think of the questions your customers have in mind and try to find ways to provide that information in an easy-to-consume manner.

When your customers discover your APIs, it might be through channels such as social media such as YouTube, Twitter, and so on; Google Search; events; newsletters; white papers; and so on that lead them to your developer landing page where they can start to learn about your APIs. These channels are the first touch point between you and your customer and are also at the top of your customer funnel. Chances are that not all customers who land on your developer landing page will end up learning more about your API offering, where they would consider pricing, read through the documentation, and so on. The number of people who would spend time on your developer landing page and learn about integrating with your APIs is the number of customers who show more intent to integrate with your APIs. You can measure the percentage of these customers, and that percentage is considered the conversion rate of customers at the discovery stage. In the same way, you can figure out conversion rates for each stage of the customer journey and for each touch point along the way.

These conversion rates are important because they help you figure out how effective and useful the tools and information you give your customers on their journey are. In the following section, you will learn how to evaluate these touch points as points of friction or conversion.

Identifying the points of friction and conversion

All the touch points that a customer experiences as they go through the user journey are components of your API experience. The API experience includes everything from developer landing pages, API references, blogs, video tutorials, as well as support channels. As you analyze the customer journey, the question you must ask is which touch points are helping your customers and which ones could be improved.

For example, if you have 100,000 views on your YouTube tutorial on getting started with your APIs but only a few hundred users are clicking on the link to your developer landing page to sign up and get started, there might be an opportunity to improve the content or place the link in a more visible position to help discovery. Your ability to measure any component or content is dependent on having clear call to actions at each step that you can track.

Your user experience must be designed in a way that there is an unambiguous call to action on each page that your user views and a finite set of actions that the user takes to move to the next stage of integration. This ensures that your users have clarity on how to move ahead in the process and that pages are not overloaded with paths that they can take.

The conversion rates for each stage of the customer journey are known as the conversion funnel for your customers. When a user moves from the first step of discovery of the product to a significant stage in the customer journey in terms of business impact, you can set that stage as the activation rate. For example, if from 100,000 views on YouTube, 800 users were able to sign up on the developer portal to get their credentials and make their first API call, then those 800 users have successfully activated their accounts and you have an activation rate of 0.008% from YouTube. In this case, the event of the first API call is significant in the user journey.

You can also track which pages have led to the most support tickets being created to date. It is expected that pricing pages will lead to a number of support requests where customers ask clarifying questions regarding pricing. However, this is not always the case. If the pricing is far too high, customers might leave the page without further inquiry.

You can also run surveys on various components of your API experience to get more quantitative and qualitative feedback from customers while they are on the pages. Surveys ensure that the feedback you receive is immediate and not anecdotal. In the end, finding points of friction will help you figure out exactly which parts of your API experience need to be improved first. You could also find out which parts of your product are working and which aren't by measuring conversion and getting regular feedback from users on each experience.

Summary

In this chapter, you learned about user personas and mapping the user journey. Consumers of APIs have a view of the APIs that starts with their discovery of the APIs. Users will go through discovery, evaluation, integration, deployment, and observability stages before you can consider them to have successfully onboarded to your APIs. During this process, team members from different parts of the organization may work together to make decisions and put them into action. As the producer of APIs, it is important that you develop a deep understanding of all the different personas involved in making this buying decision.

Once you identify the various user personas for your APIs, you can work backward from the customer journey to identify their pain points, enabling you to measure exact customer pain points so that you can address them in a prioritized manner. A deep understanding of your users will empower you to communicate with your team and other stakeholders in a consistent language about customer needs.

In the following chapter, we will build on our understanding of the developer's journey and learn how to align organizational goals and stakeholder incentives using quantitative and qualitative methods of user research.

8
Customer Expectations and Goals

As the producer of an API product, your goal is to get more and more customers to successfully integrate and use your APIs regularly. From the perspective of the consumer, the goal is to be able to integrate your API quickly to meet their business needs so that they can realize the business value of using your APIs as soon as possible.

In this chapter, you will learn more about the following topics:

- Conducting qualitative and quantitative user research
- Creating user empathy maps
- Identifying customer use cases
- Identifying customer pain points
- Aligning stakeholders

Conducting qualitative and quantitative research

You can use various techniques to reach out and connect with customers to stay informed on various pain points. Quantitative user research is research that gives you numbers, while qualitative research gives you information that is harder to fit into a calculation. What kind of research you do depends a lot on your research goals and what kind of data will help you understand your users' needs the best.

Don't think that either type of research is less important than the other. Both can give you valuable information that can help guide your design process and lead to great results.

For quantitative research, you use different types of user testing to collect and analyze data that is objective and can be measured. Quantitative data almost always consists of numbers, and its analysis is based on statistics, math, and computers. As the name suggests, the goal of quantitative user research is to get measurable results.

Analytics is an effective way to get a lot of quantitative data for UX design. Using analytics, you can track things such as how many people visit a page, how many people stay on the page, and how many people buy something.

User testing sessions are also a great place to get quantitative information. You can get quantitative data from user testing, such as the time it takes to finish a task, the number of mouse clicks, the number of mistakes, and the success rate.

Since quantitative user research is objective, the data that comes out of it is less likely to be affected by human bias. This is because it's harder to lead participants to a certain result and the study conditions are well defined, strict, and controlled. Quantitative data is also usually easy to collect, quick to analyze, and easy to see in pie charts, bar graphs, and so on. Customers may also prefer to see hard numbers and find it easier to connect them to their KPIs as a way to justify spending money on improvements for the future.

Qualitative research gives you a deeper look at your users and often shows things that quantitative data can't. Qualitative testing uses a "*think aloud*" method that lets you walk in your customers' shoes and learn about how they use your API product in their own environment and their responses to and frustrations with it. Qualitative data lets you make good decisions for your users without guessing about what caused what. Getting this kind of evidence, which is based on empathy and emotion, can make it easier for stakeholders to invest in making changes to the product.

In the following sections, we'll look at some common qualitative and quantitative ways to do research on our users.

Qualitative research methods

Here are some of the most common ways to do qualitative research for user research:

- **User interviews**: User interviews follow a structured method in which the interviewer comes up with a list of topics to talk about, writes down what is said during the interview, and then analyzes the conversation systematically after the interview.
- **Focus groups**: Focus groups are used to understand what customers think about new ideas or concepts and how they feel about them. Most of the time, they are used during the design phase and the early stages of research to find out what customers think. Focus groups are also helpful after the product has been released because they help users share their thoughts on a product that already works. A focus group should contain between 8 and 10 people. It is also suggested to run three or four separate focus groups to get a good mix of ideas and points of view.
- **Diary studies**: In a diary study, the people taking part report their own data longitudinally, which means over a long period of time – anywhere from a few days to a month or more. The participants in a study are asked to keep a diary and write down specific information about the activities under study for a set period. Participants are sometimes reminded to fill out their diaries from time to time (for example, via a notification received daily or at select times during the day).

- **User shadowing**: Shadow sessions are a user research method that focuses on observing and documenting the actions of a user as they interact with a product or service at a specific moment in time. They are typically used to study the user experience and identify areas for improvement. Shadow sessions are a quick way to get a glimpse of how a user interacts with a product and are typically conducted in a lab setting or remotely through screen-sharing. User shadowing is when you watch a person use a product or service for user research purposes. The goal is to collect rich contextual data (usually written or spoken) that supports a user problem and answers the *what* and *why* of your research through real-life experiences. When you watch a user interact with a product *face to face*, you can learn many things about it that you wouldn't have known otherwise.
- **Ethnographic research**: Ethnographic research is a qualitative research method that is used to study the culture, behavior, and beliefs of a particular group or community. It typically involves observing people over an extended period and collecting data through methods such as observation, interviews, and participant observation. The goal is to understand how people live, think, and experience the world around them. Both ethnographic research and shadow sessions are valuable methods for understanding user behavior, but they are used in different ways. Ethnographic research is used to gain a deep understanding of a culture or community, while shadow sessions are used to understand how a specific product or service is being used.

You may not have to do all of these, but you should be able to select different methods based on what kind of research you conduct.

Quantitative research methods

You can get a better understanding of how people use your site with a few common quantitative methods. In contrast to qualitative methods, quantitative methods can be measured with tools, take less time over time, and can be used over and over again:

- **Net Promoter Score (NPS)**: The NPS is a metric that measures loyalty. It uses a scale from 0 (not likely) to 10 (very likely) to estimate how likely someone is to recommend a company, product, or service to a friend or co-worker.

  Based on how customers answer on the 11-point scale, they are put into one of three groups:

  - Those who respond with a 9 or 10 are **promoters**
  - **Passives** are people who answer with a 7 or 8
  - Those who answer between 0 and 6 are **detractors**

  The NPS is found by taking the percentage of detractors and subtracting it from the percentage of promoters. Your NPS can range from -100 to 100. A percentage should never be used to show the calculated score. Passives are never taken into account when figuring out the NPS.

- **Customer Satisfaction Score** (**CSAT**): Customer feedback is used to measure the CSAT. One or more versions of this question are used to find out the following for each customer: *How satisfied were you with the [goods/services]?* A 1–5 scale is used by those who answer:
    - Very unsatisfied
    - Unsatisfied
    - Neutral
    - Satisfied
    - Very satisfied

  You can average results to get a CSAT, but CSATs are usually given on a percentage scale.
- **Funnel analysis**: Funnel metrics are a set of metrics that help businesses measure the effectiveness of their sales funnels. These metrics also help companies figure out what's wrong with their marketing and sales funnels and fix them, so measuring the metrics of your funnel is a great way to figure out how to get more customers.
- **Heatmaps**: Heatmaps are graphical representations of data that use color to show the density of data points in a specific area. They are often used to visualize data that has a geographic component, such as website traffic or user interactions. Heatmaps are a useful tool for visualizing data, allowing us to identify patterns and areas for improvement in the user experience, usage, and performance of the API. For example, a heatmap of a website that utilizes an API could show the areas of the site where users spend the most time. This information can be used to identify which pages and features are the most popular and where users experience the most difficulty. Heatmaps can also be used to analyze the usage of specific features within an API. For example, a heatmap of an e-commerce site's checkout process could show which steps in the process cause the most friction for users. This information can be used to improve the checkout process and make it more user-friendly. Additionally, heatmaps can be used to analyze the performance of an API by showing the response times for different regions. This can help to identify potential bottlenecks and areas in which performance can be improved.
- **Cohort analysis**: Cohort analysis is the method of analyzing how different groups of people act. Each group is called a **cohort** because it has something in common. By putting people into cohorts, you can look at many different parts of a business, such as how to get customers, keep them, and make money. It can also tell you which customers you should focus on, which forms of communication work best, and which parts of the experience need more resources to be improved.

Combination of qualitative and quantitative methods

Even though the techniques we've seen so far can be categorized as either qualitative or quantitative, there are research methods that combine the two to give you a more complete picture of the customer:

- **Surveys**: A survey is a set of questions you use to collect data from your target audience. Surveys are appealing because they reach a large number of people quickly and easily without spending a lot of money. You can ask about almost anything, and it's easy to add up the answers, but this is exactly why surveys are so dangerous; they make it seem like they are easy when they aren't. Even though they're easy and cheap to use, you shouldn't let that make you think you can cut corners. To get valid and useful information from a survey, you still need to stick to best practices.
- **A/B testing (split testing)**: A/B testing is a quantitative way to find the best **Call-to-Action (CTA)**, copy, image, or any other variable. To start A/B testing, make two or more versions of a single element, randomly split your user group in half, and see which version works better. When making a digital product, there are many choices to be made. For example, which font is easiest to read? Which CTA gets more people to act? Having so many options make it hard for designers to decide what to do. Best practices and your gut are good places to start, but they won't get you very far in business, and bad design choices can hurt your revenue stream. What should you do then? Base all your UX decisions on solid data using A/B testing.
- **Card sorting**: In the past, card sorting was used to evaluate information architecture. This method was mostly used to set up the navigation and structure of websites, but this simple-looking method can be used in a lot more ways than people think. IBM design researchers also use it for things such as telling stories, creating multi-level hierarchies, and figuring out what's most important.

  In card sorting, one idea is written on each card by the researcher. These can be index cards if you're doing the exercise in person, or you can use digital cards. Cards are then placed in a random order to avoid sorting bias and participants sort cards into categories that make the most sense to them. This often results in a two-level hierarchy of a category and its parts.

Qualitative and quantitative user research are both needed to design products and experiences that really meet the needs and goals of users. When you do both quantitative and qualitative research, you can come up with hypotheses and ways to test them. When you only do one type of research, you often end up with unanswered questions and metrics that are vague or wrong. When used together, quantitative data answers questions such as "*what?*," "*how many?*," and "*how much?*" while qualitative data answers questions such as "*why?*".

Even though quantitative and qualitative user research methods have different goals, they work well together and give you a fuller, more complete picture of how well their product designs are doing. In *Chapter 14*, you'll learn how to understand the results of research and get rid of any biases that could affect the quality of your conclusions.

The end goal of the research is to learn more about customers' pain points so that you can use that information to make decisions about your products. In the following section, you will learn how the results of the qualitative and quantitative research help identify customer pain points.

Creating user empathy maps

As a product manager, it's your job to look out for the user's best interests. However, in order to do so, you will need methodologies that allow you to document your understanding of the customers, as well as help your colleagues understand them. Empathy maps are used a lot in both the agile community and wider design communities. They are an important and powerful tool in both cases. With user empathy being such a central driver of product strategy, it is not the responsibility of a single person, and it is also not accomplished in one go. User empathy needs to be built into every aspect of the product experience and every decision that is made during product development.

Design a user research strategy that allows you to survey and interview customers on a variety of scales and use cases. As we learned in *Chapter 7*, the most important place to start with your user research strategy is by making a customer journey map to see how customers find, evaluate, and use your APIs. Segment your audience so that you have a clear understanding of the target segment as you start your user research. Within your target segment, you can identify three to five personas with varying degrees of influence on making the buying decision for your product. For each persona, you can put together a user empathy map that will help you document the factors that these personas consider in their decision-making process.

A user empathy map is a way of documenting and communicating your understanding of user personas and their decision-making processes in a way that is consumable by other members of your team. As you research the users and study their user journeys, share the insights with the entire team to build a shared understanding of the customer personas and their needs. This makes user knowledge public so that making decisions is easier and the whole team has a better idea of what users want.

This will allow you to build a repository of user empathy maps for various personas that you can bring up as you scope features specific to these personas, as well as when you speak to stakeholders who might already have a deep understanding of these personas. The following figure shows the template for a user empathy map. You can use this template to make notes in each of these sections of **SAYS**, **THINKS**, **DOES**, and **FEELS** for each persona that you identify to create a user empathy map:

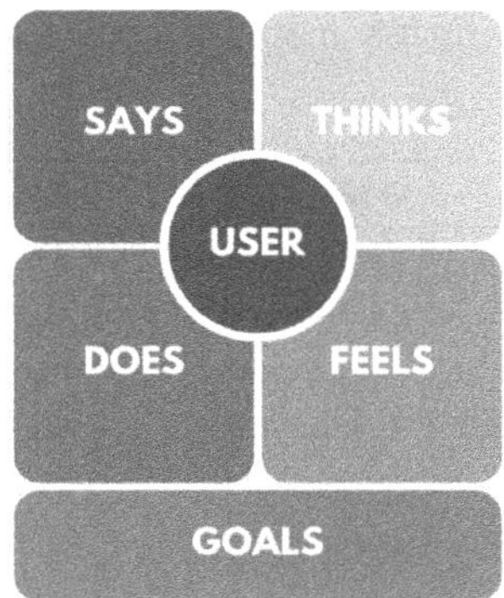

Figure 8.1 – User empathy map template

Empathy maps are made with the help of qualitative inputs such as user interviews, field studies, diary studies, listening sessions, or qualitative surveys. Empathy maps enable us to empathize with our end users. They also help us remove bias from designs and align the team around user understanding. User empathy helps you find out about consumer needs you might not have known about and figure out what makes people act the way they do.

When you speak to customers, you can try to identify their persona types and map their needs to your current understanding of a persona. You may notice that user empathy maps evolve as new tools and features enter the market. This is expected. User research is an ongoing practice, and you should constantly update your empathy maps to reflect the evolving needs of your customers.

Imagine you are trying to create user empathy maps for a payment API – the data for the empathy map might look like the following:

Says	Thinks	Does	Feels	Goals
"I need a fast and secure payment API that can handle a high volume of transactions."	"I'm worried about the security of my transactions and whether the API can handle the volume of transactions that I need."	Research different payment APIs and compare their features, pricing, and customer reviews.	Anxious, stressed, and uncertain about the security of the transactions.	To find a fast, secure, and reliable payment API that can handle a large volume of transactions and meet business needs.
"I want an easy-to-use API that is well documented and has a user-friendly interface."	"I want to be able to easily integrate the API with my existing systems and platforms."	Look for documentation, tutorials, and customer support when evaluating different payment APIs.	Frustrated and time-pressured if the API is not easy to use and well documented	To find an API that is easy to use, well documented, and can easily be integrated with existing systems and platforms.
"I'm concerned about the cost of using the API and the return on investment."	"I want to make sure that using the API will provide a good return on investment."	Research the pricing and costs of different payment APIs and compare them to the potential return on investment.	It is uncertain whether the API will provide a good return on investment.	To find a payment API that is cost-effective and provides a good return on investment.

Table 8.1 – Empathy maps for a payment API

There is a wealth of insights available in this kind of data. You can use this data to analyze what the customer use cases and pain points are, which you will learn about in the following sections.

Identifying customer use cases

Use cases are descriptions of how a user or customer interacts with a product or service. They describe the actions that a user will take, the goals they hope to achieve, and the results they expect to see.

Deriving customer use cases from customer empathy maps is important because it helps us understand the specific needs and wants of customers, and how they will use the product or service. By understanding the customer use cases, the product or service can be designed and developed to meet the needs of customers better.

When creating customer empathy maps, you can understand what the customer says, thinks, does, and feels and their goals, and use this information to create use cases that reflect the customer's needs. For example, if the customer needs a fast, secure, and reliable payment API that can handle a high volume of transactions, the API provider can create a use case that describes how the API will handle high-volume transactions securely and quickly.

Using the example of a payment API, here are some examples of customer use cases for a payment API using the information provided in the previous empathy map:

- **High-volume transactions**: Somebody who needs to process a large volume of transactions, such as on an e-commerce site, would use the API to securely and quickly process payments. They would want the API to be able to handle a high volume of transactions and ensure the security of their customers' sensitive information.
- **Easy integration**: A customer who wants to integrate the API with their existing systems and platforms, such as a point-of-sale system or an inventory management system, would use the API to easily and seamlessly integrate payments into their existing processes. They would want the API to be well documented and easy to use and have a user-friendly interface.
- **Cost-effectiveness**: A customer who is concerned about the cost of using the API and the return on investment, such as a small business, would use the API to ensure that the cost of using it is justified by the potential return on investment. They would want the API to be cost-effective and provide a good return on investment.
- **Mobile payments**: A customer who wants to process payments through mobile devices, such as a mobile app, would use the API to process payments through the mobile app, and they would want the API to be secure and easy to use.
- **Recurring payments**: A customer who wants to process recurring payments, such as a subscription-based service, would use the API to set up and manage recurring payments, and they would want the API to be reliable and easy to set up.
- **Refunds and cancelations**: A customer who wants to process refunds and cancelations would use the API to process these actions, and they would want the API to be flexible and easy to use.

Overall, creating customer use cases from customer empathy maps is an important step in understanding customer needs and creating a product or service that is tailored to meet those needs. This will help ensure that the product or service provides value to customers and that it's more likely to be adopted and used.

Additionally, creating use cases from customer empathy maps can also help to identify potential pain points and areas for improvement. In the next section, you will learn more about identifying customer pain points and how you can develop solutions that address pain points and improve the customer experience.

Identifying customer pain points

While use cases describe the actions that a user will take, the goals they hope to achieve, and the results they expect to see when using a product or service, customer pain points are the areas of difficulty or dissatisfaction that customers experience when using a product or service. They are the areas in which the product or service does not meet the customers' needs or that cause frustration or inconvenience. They are the areas where the product or service can be improved to better meet the customers' needs.

Creating use cases from customer empathy maps can help to develop an understanding of customer pain points because customer empathy maps provide a detailed understanding of customer needs, wants, and issues. By understanding the specific use cases that apply to customers, a product or service can be designed to better meet those needs and address the pain points.

For example, if a customer empathy map indicates that customers want a fast, secure, and reliable payment API that can handle a high volume of transactions, and the use case is "process a high volume of transactions," the API provider can focus on improving the performance and security of the API to better meet this use case and address the customer pain point of slow transaction processing or security concerns.

In *Chapter 7*, you learned about mapping the user journey and identifying customer personas, and creating user empathy maps should give you a good understanding of your customers' needs. Trace the customer journey and try to identify pain points at various steps of the journey to find the most impactful opportunities for improvement.

Customer pain points can be categorized into four main categories: financial, product, support, and process:

- **Financial pain points**: These are pain points related to the cost of the product or service, such as high prices, hidden fees, or unexpected charges. They also include concerns about the return on investment, such as whether the cost of the product or service is justified by the benefits it provides. Examples of financial pain points include the following:
    - **High transaction fees**: Customers may feel that the transaction fees are too high and not competitive compared to other payment providers

    - **Hidden fees**: Customers may be unaware of additional fees that are not clearly disclosed, such as monthly maintenance fees, chargeback fees, or fees for refunds
    - **Unexpected charges**: Customers may experience unexpected charges that they were not aware of, such as fees for dormant accounts or inactivity

- **Product pain points**: These are pain points related to the features and functionality of the product or service, such as poor performance, lack of features, or poor user experience. They also include concerns about the reliability and security of the product or service. Examples of product pain points include the following:
    - **Slow transaction processing**: Customers may experience delays in the processing of their transactions, which can lead to frustration and dissatisfaction
    - **A lack of security**: Customers may be concerned about the security of their sensitive information, such as credit card numbers or personal information
    - **Limited payment options**: Customers may be limited in the types of payment methods they can use, which can lead to inconvenience and frustration
- **Support pain points**: These are pain points related to the support provided by the company, such as poor customer service, a lack of documentation, or difficulty getting help when needed. Support pain points include the following:
    - **Poor customer service**: Customers may have difficulty getting help when they need it, or may not be satisfied with the quality of the customer service that they receive
    - **Limited documentation**: Customers may have difficulty finding the information they need to use the API, or may find the documentation to be poorly written or hard to understand
    - **Difficulty getting help**: Customers may have difficulty getting help when they need it, or may not be satisfied with the quality of the help they receive
- **Process pain points**: These are pain points related to the process of using the product or service, such as a difficult sign-up process, confusing navigation, or long wait times:
    - **A difficult sign-up process**: Customers may have difficulty signing up to use the API, such as a long and complicated registration process, or may be required to provide excessive personal information
    - **Confusing navigation**: Customers may find the API difficult to navigate, or may have difficulty finding the information they need
    - **Long wait times**: Customers may experience long wait times when trying to use the API, such as when trying to process a transaction or get help

By categorizing customer pain points into these four areas, it can be easier for a company to understand where the pain points lie and focus their efforts on addressing the specific issues that are causing the most grief for their customers. This will help the company to improve the customer experience and increase customer satisfaction.

You can also bucket the pain points you see across the various customer personas you study to see which types of pain points are most common for your customers. This will allow you to group customer personas that you could then target with new features and fixes. Once you have identified customer pain points and the personas impacted by these pain points, you can align this with your business strategy to determine which pain points would be most impactful to address in the immediate, medium, and long term.

When you have a deeper understanding of customer pain points, you will also be able to find the stakeholders who would be the best partners to help you solve these problems. In the next section, you'll learn how to make sure that stakeholders are on board with the projects you decide to prioritize based on your user research.

Aligning stakeholders

You learned about the stakeholders that product managers partner with in *Chapter 2*. When you are thinking of offering better tools or experiences for your customers, several stakeholders will also be invested in your goals to ultimately drive success for the organization. These stakeholders include sales teams, who are incentivized to gain more customers, support teams, who are incentivized to provide better service to the customers, and so on. Legal, compliance, and security teams are often very crucial partners in building enterprise products; this is because making compliant products or offering security features can be key differentiators for APIs.

At this point, various stakeholders are involved in various aspects of the customer journey and are working to improve the customer experience. You can identify the right stakeholders to work with for every feature to identify the downstream impact of the product changes you plan to make. For example, if you are fixing a prevalent bug that is going to change the number of support requests received by the support team, it would be beneficial for the support team to be involved in the planning of this feature so that they can anticipate and confirm the results. Similarly, when you plan to ship new features for a specific target audience, sales and marketing teams would need to be in the know so that they can plan sales and marketing activities around that feature.

You can help your product succeed by figuring out who is important, keeping in touch with them, and working with them effectively.

Summary

In *Chapter 7*, you learned how to map the developer journey and figure out who your customers are. In this chapter, you learned how to go a step further and create customer empathy maps and conduct qualitative and quantitative research to identify customer use cases and pain points. Using the framework in this chapter, you can create artifacts such as empathy maps, as well as gathering the data you need to align with your stakeholders.

When building complex products, it is often easy to lose sight of the end customer. You will need to align your team and all the stakeholders with the vision of driving customer value and driving success. Establishing a method of continuous user research that lets you learn more about different customer segments and personas and gain a deeper understanding of the pain points of these personas will be key to making sure that everyone on the team has the same understanding of the customer.

Now that you know what your customers want and what bothers them, you will learn how these problems can be solved using different developer experience components. In the next chapter, you'll learn how different parts of the developer experience, such as API documentation and tools for development, testing, and observability, can be used to address customer pain points and use cases.

9

Components of API Experience

API experience encompasses all documentation and tooling made available to customers to assist them in learning about and integrating with your APIs. Aside from the tools, the API experience also includes customer support channels and developer relations efforts, which have a big impact on how customers use your APIs.

As you learned in *Chapter 7, Walking in the Customer's Shoes*, each step of the customer journey can be mapped to multiple components of the API experience. In this chapter, we will look at which industry standards exist for API experiences and what the key components that make up an industry-standard API experience are. This will include the following topics:

- Industry standards for API experience
- Creating API documentation
- Providing developer tools
- Instrumenting support mechanisms

By the end of this chapter, you'll have a toolbox of API experience components that customers look for when evaluating APIs for integration.

Industry standards for API experience

Over the years, many companies building API products have organically converged on a set of standardized documentation patterns and tools that they provide to their customers to learn about their APIs. Although there isn't a specific list or guide to building API experience components, there are best practices that have been established by the leaders in the space.

API documentation starts with an API reference, which is often made directly from API specifications. There are additional components such as a sandbox, a **command-line interface** (**CLI**), **software development kits** (**SDKs**), Postman Collections, and so on to aid in the developer's journey. In addition to tools that help developers get started with using your APIs, there are also tools that are focused on existing customers who might be looking at the API documentation to diagnose a problem with their integration or make enhancements to their existing integration.

There is an opportunity to be innovative and introduce new tools and techniques to improve the user experience for APIs, and this can be a great opportunity for new product managers to bring new perspectives. In the following section, you will learn about the key components of API experiences and the value they deliver to customers.

Creating API documentation

The first thing a customer needs to start working with APIs is a set of documentation that introduces and educates them about the features and functionality of the APIs. Once they are ready to start exploring the APIs further, they should be able to generate API credentials they can use to start making their first API calls. To make this process self-serve, the documentation and developer portal must be designed in a thoughtful manner. In this section, you will learn about the various components that are available on a developer portal to aid a developer's journey from discovery to consideration, integration and testing, launch, and operational support, such as API references, the developer portal, API status pages, and the changelog, as well as integration guides and tutorials.

API references

API references are documentation that provides information about the methods, parameters, and other elements of an API. This is usually provided in the form of a technical reference manual or a set of online documentation pages, and it is intended to help developers understand how to use the API and integrate it into their own applications.

API reference documentation is important because it helps developers understand how the API works and how to use it effectively. By providing detailed information about the API's methods, parameters, and other elements, the documentation can help developers understand how to use the API to achieve specific goals and make it more likely that they will be successful in integrating the API into their own applications. In addition, API reference documentation can also help developers troubleshoot problems and find solutions when they encounter issues while using the API.

API reference documentation typically consists of a set of pages that describe the various elements of the API. Here is a brief description of each type of page:

- **Overview page**: The overview page is the starting point of the API reference documentation. This page should include information on pricing and rate limits, along with the maturity of the API, to enable customer decision-making.

- **Interface pages**: An interface page is a page that describes a specific interface in the API. An interface is a collection of related methods and properties that define a set of behaviors. The interface page lists all of the methods and properties that are part of the interface and provides links to the individual documentation pages for each method and property.
- **Constructor pages**: A constructor page is a page that describes a specific constructor in the API. A constructor is a special type of function that is used to create new objects. The constructor page describes the parameters that the constructor takes and the properties and methods that are available on the objects that it creates.
- **Method pages**: A method page is a page that describes a specific method in the API. A method is a function that is associated with an object or interface. The method page describes the parameters that the method takes, the return value of the method, and provides examples of how to use the method.
- **Property pages**: A property page is a page that describes a specific property in the API. A property is a value that is associated with an object or interface. The property page describes the type of the property, the default value of the property, and any constraints on the property value.
- **Event pages**: An event page is a page that describes a specific event in the API. An event is a notification that is triggered by certain actions or conditions. The event page describes the circumstances in which the event is triggered, the data that is provided with the event, and any methods or properties that are related to the event.
- **Tutorials**: Tutorials are step-by-step guides that show developers how to use the API to build a specific type of application or perform a specific task. These pages can include code examples and detailed explanations of the concepts and techniques involved.
- **How-to guides**: How-to guides are shorter, task-focused articles that provide information on how to perform specific tasks with the API. These pages might include code examples and tips for troubleshooting common issues.
- **Glossary**: A glossary is a list of terms and definitions that are used in the API documentation. This can be helpful for developers who are unfamiliar with the API or certain concepts that are used in the documentation.
- **FAQ**: A **frequently asked questions** (**FAQ**) page is a list of common questions and answers about the API. This can be a useful resource for developers who have questions about how to use the API or troubleshoot issues.
- **API changelog**: An API changelog is a list of changes and updates that have been made to the API over time. This can include new features, bug fixes, and other changes that affect how the API works.
- **API status page**: An API status page is a page that provides information on the current status of the API, including any planned or unplanned downtime or maintenance. This can be helpful for developers who need to know whether the API is available or whether they should expect any disruptions to their applications.

The purpose of these pages is to provide detailed documentation for each element of the API so that developers can understand how to use the API and build applications with it.

When making API references, there are two main things to keep in mind: **information architecture** (**IA**) and **user experience design** (**UXD**). IA helps organize the huge amount of information in API references, but it's the UXD of API reference pages that makes this information usable. In the following sections, you will learn about IA and UXD key considerations when building API reference documentation.

IA

IA is the practice of organizing and labeling the content of a website or application in a way that makes it easy for users to find and use the information they need. This includes organizing the content into logical hierarchies and categories and using clear and descriptive labels for the content and navigation elements.

IA is a key part of making good API reference documentation because it helps developers understand and use the API. A well-designed API reference should have a clear and logical structure, with information organized into categories and subcategories that reflect the way the API is designed and used. This makes it easier for developers to find the information they need and to understand how the different parts of the API fit together.

There are several factors to consider when designing the IA for API reference documentation:

Organization: The documentation should be organized in a logical and intuitive way so that developers can easily find the information they are looking for. This might include organizing the documentation by topic, resource type, or by the API's functional areas.

Navigation: The documentation should include clear and consistent navigation aids, such as a table of contents or a search function, to help developers find their way around the documentation.

Cross-referencing: The documentation should include cross-references to related topics or concepts, to help developers understand the context and relationships between different parts of the API.

Consistency: The documentation should use a consistent style and formatting across all pages, to make it easier for developers to read and understand the information.

Scannability: The documentation should be written in a way that makes it easy for developers to scan and quickly understand the key points. This might include using headings, lists, and code examples to break up the text and highlight important information.

Accessibility: The documentation should be accessible to developers with disabilities, and should be optimized for both desktop and mobile devices.

Maintainability: The documentation should be easy to maintain and update over time, as the API evolves and changes. This might include using a documentation generator or other tools to automate the documentation process.

Use descriptive and meaningful names: The names you choose for your documentation elements (for example, methods, properties, events) should be descriptive and meaningful so that developers can easily understand what they do. Avoid using abbreviations or acronyms unless they are widely understood.

Use a consistent naming style: Choose a naming style and stick to it consistently throughout the documentation. For example, you might choose to use camelCase for method and property names, or PascalCase for class names.

Use plural names for collections: If an element represents a collection of items (for example, a list of users), consider using a plural name (for example, "users") to indicate this.

Use suffixes to indicate the element type: Consider using suffixes (for example, `ID` for identifiers, `URL` for URLs) to indicate the type of an element. This can help developers understand the expected format or type of a value.

Use prefixes to indicate visibility: Consider using prefixes (for example, `private_` for private elements, `protected_` for protected elements) to indicate the visibility of an element. This can help developers understand which elements are intended to be used within the API and which are intended for internal use only.

Naming conventions can be an important part of the IA for API reference documentation, as they help to establish a consistent and intuitive structure for the documentation. IA is also subject to review from compliance, branding, and legal teams to ensure consistency and correctness of the terms used.

The way that the information on a website or application is organized and labeled plays a critical role in determining how easy it is for users to find and use the information they need. Even the most visually appealing websites might not be able to meet their users' needs if they don't have a good IA. On the other hand, even websites that look simple or plain can be good if they are set up in a way that makes sense and is easy to use. In the end, a good IA is one of the most important parts of making a good user experience.

UXD considerations

In addition to a good IA, a good UXD is also important for successful API reference documentation because it can make the documentation easier to use and navigate. This can be especially important for API reference documentation, which can be complex and technical.

By designing the documentation with the user in mind, and considering things such as layout, navigation, and readability, developers can more easily find the information they need to understand how to use the API. When the documentation is well designed, developers and decision-makers are able to find the information they need quickly and easily, and they are more likely to use the documentation as a resource when working with the API.

There are several key UXD best practices to keep in mind when creating API reference documentation:

Navigation: The documentation should be organized in a way that is logical and intuitive, and should include clear and descriptive labels for the different sections and categories. This makes it easier for users to find the information they need.

Search functionality: The documentation should include a search function that allows users to quickly and easily find specific information or keywords.

Concise and accurate information: The information in the documentation should be clear, concise, and accurate. It should provide detailed explanations of how the API works and how it can be used, but should avoid using unnecessary jargon or technical language.

Code samples and examples: The documentation should include code samples and examples to illustrate how the API can be used in practice. These can be especially helpful for developers who are learning how to use the API.

Responsive design: The documentation should be designed to be responsive so that it is easy to use on different devices and screen sizes.

Error messages: If the API returns error messages, the documentation should provide clear and concise explanations of what the errors mean and how they can be resolved.

Page length: The length of each page of documentation should be appropriate for the information it contains. Too much information on a single page can be overwhelming and difficult to navigate, while too little information can make it difficult for users to understand how the API works and how to use it effectively.

Scrolling: The amount of scrolling required to access all of the information on a page should be minimized. Users should be able to access all important information on a page without having to scroll excessively.

White space: The use of white space, or the empty space on a page, can help to break up large blocks of text and make the documentation easier to read and navigate.

Visual design: The design of the documentation should be visually appealing and consistent with the overall look and feel of the API. The use of images, graphics, and other visual elements can help to make the documentation more engaging and easier to understand.

By considering these UXD elements when creating API reference documentation, it is possible to create a resource that is easy to use and understand, which can help increase the adoption and usage of the API.

Developer portal

An API developer portal is a website that provides developers with the information and resources they need to effectively use and integrate with a company's API. It serves as a one-stop shop for developers to access documentation, tutorials, code samples, and other resources that help them to understand and use the API.

It is common to use a developer's <domain name> as the URL for the developer portal. The following screenshot shows the developer portal for Google (https://developers.google.com/) where developers can explore all the developer tools offered by Google:

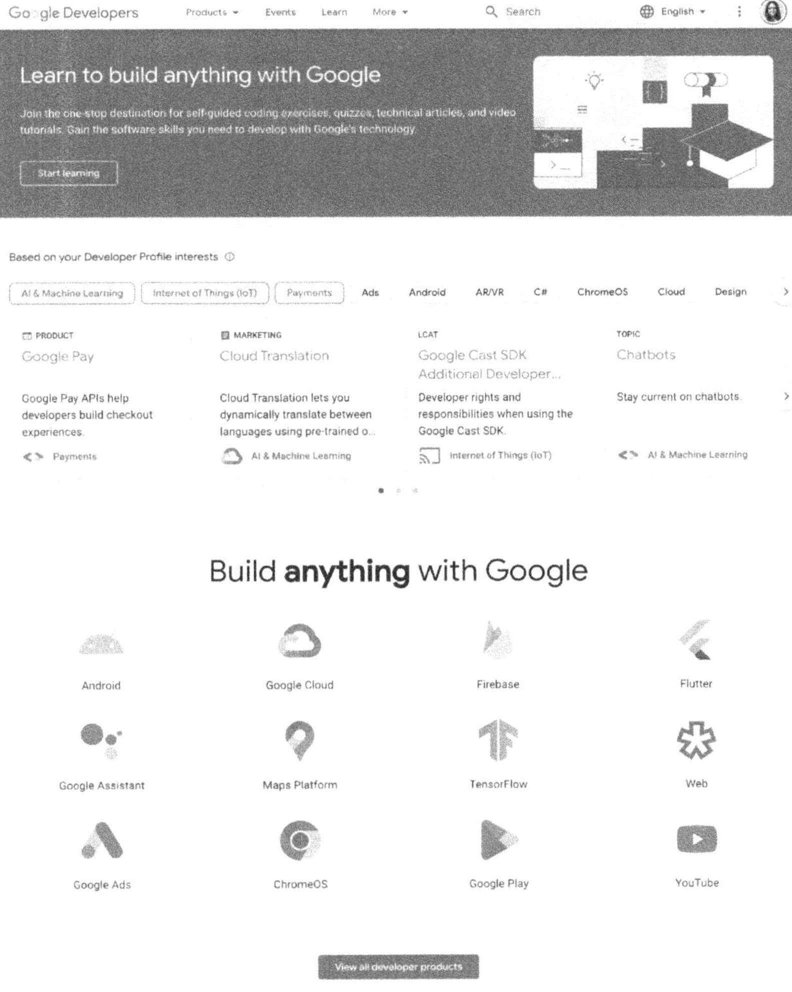

Figure 9.1 – Google developer portal landing page

The Google developer portal also offers a developer expert directory that allows experts on Google developer tools to connect with organizations that need them:

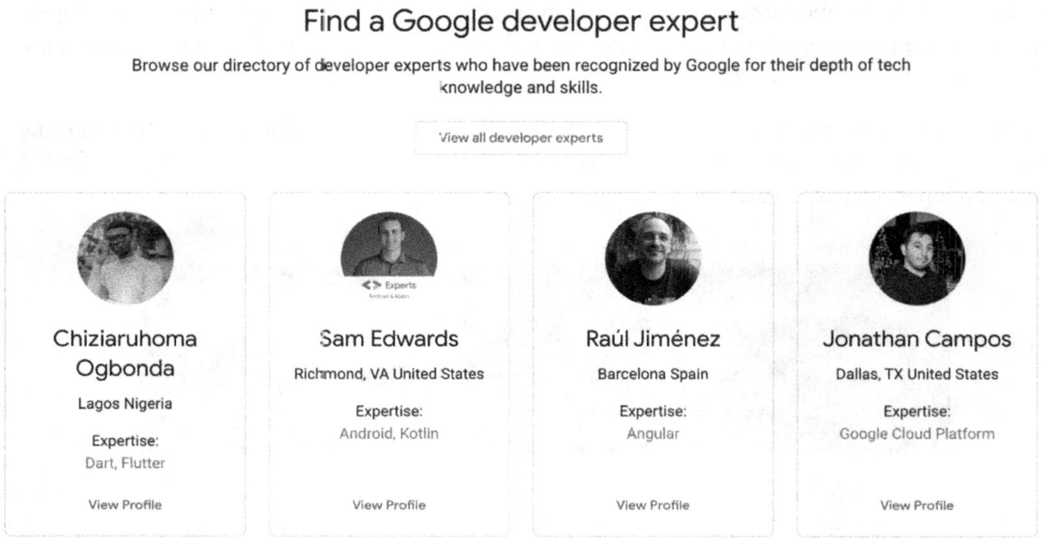

Find a Google developer expert

Browse our directory of developer experts who have been recognized by Google for their depth of tech knowledge and skills.

View all developer experts

Chiziaruhoma Ogbonda	Sam Edwards	Raúl Jiménez	Jonathan Campos
Lagos Nigeria	Richmond, VA United States	Barcelona Spain	Dallas, TX United States
Expertise: Dart, Flutter	Expertise: Android, Kotlin	Expertise: Angular	Expertise: Google Cloud Platform
View Profile	View Profile	View Profile	View Profile

Figure 9.2 – Google developer expert directory showcased on Google's developer portal

Google offers an API console where developers can manage their applications and credentials, as shown in the following screenshot:

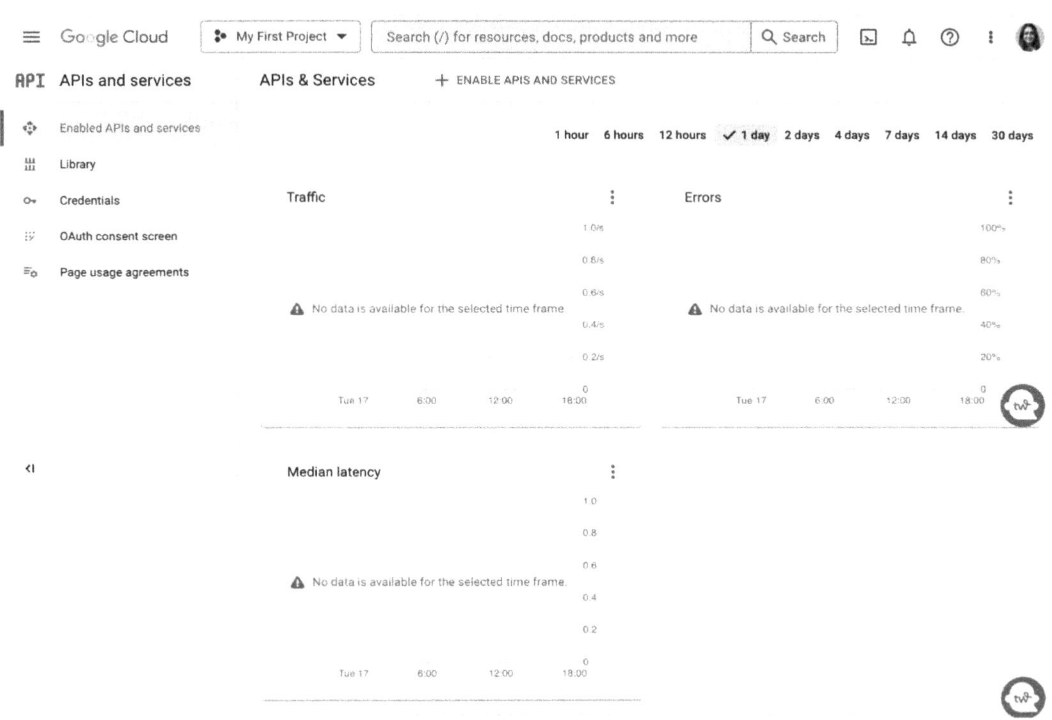

Figure 9.3 – Google Cloud API console logged-in experience showing all the tools and product offerings

API developer portals are essential for any organization that offers an API, as they help to increase the adoption of the API by making it easy for developers to understand, use, and integrate it into their applications. A developer-friendly portal can also help to build a strong developer community and foster collaboration.

API credentials

API credentials typically consist of two main components: an API key and a secret key. Let's look at these in more detail:

- **API key**: This is a **unique identifier** (**UID**) that is used to authenticate requests to the API. The API key is typically included in the headers of each request and is used by the API server to identify the calling client.
- **Secret key**: This is a private key that is used to secure communication between the client and the API server. The secret key is typically used to sign requests to the API and is used by the API server to verify the authenticity of the request.

API credentials can also include additional information such as an **Open Authorization** (**OAuth**) token, **JSON Web Token** (**JWT**) token, or other authentication information that may be required to access the API. Some APIs may also require additional parameters such as a client ID, username, or password to be provided along with the API key and secret key. It depends on the API and the security requirements of the API.

In order to get started with using your APIs, the developers need a way to get their API credentials. The developer portal is a logged-in experience on your developer site where customers can generate and retrieve their API credentials. Although there are public APIs (such as the Google Books API) that do not require you to have any credentials to access them, most paid as well as free APIs require you to create access tokens for security purposes.

API credentials assist users in protecting their applications and avoiding exposing a complete ecosystem to dangerous and unauthorized access. Because APIs are protocols and technologies that allow apps to interact with one another, a lack of sufficient credentials to enable access can result in enormous data breaches.

API credentials are UIDs that must be added to code before making an API call. When building an API, it is critical to provide the necessary credentials to define the type of data that each user can access. Although the majority of apps and websites have publicly accessible APIs, the great majority of such platforms require authentication before any request is fulfilled.

Creating numerous API credentials is a secure way of granting access to others with whom you collaborate. You keep the default access status, while the others only access the information they require. Different APIs offer different authorization credentials, but common APIs include credential options such as the following:

- **Name**: This is the first primary credential you'll be granted. You can add extra names for your partners or affiliates to utilize.
- **API password**: This serves as your identification when requesting a token or making a call. It is generated automatically when you sign in, but you can make a new one with alphanumeric characters and no spaces.
- **Auth token**: In order to access private user data, you must include an Auth token with your request. To receive a token, the app first provides a client ID and perhaps a secret phrase. Auth credentials for service accounts, online applications, and installation applications can be produced.
- **API keys**: If a request does not contain a token, it must include an API key. The key identifies the project making the call and provides API access, reports, and quota. You must have an API key to authenticate your request in order to connect to a project or application.
- **Status**: The status of a credential indicates the degree of membership by indicating whether it is active or inactive.
- **Actions**: You can use this to either change or delete your API credentials.

In addition to getting API keys to get started with making API calls, the developer portal also provides key account-level functionality such as billing and payment information. APIs are usually priced based on usage, and even though a number of companies enable developers to try out the APIs for free, there are thresholds beyond which there are charges that can be accessed using billing pages in the developer portal.

One of the key items of interest for administrative and building applications is error logs. Developer portals provide error logs that enable customers to see a detailed log of errors produced by their applications. These should be presented in an accessible and organized way for effective debugging and analysis so that customers can improve their integration and resolve issues effectively.

Account-specific notifications can also be delivered to customers using the developer portal. Developer portals can also be designed to have multiple roles and permissions. In large organizations, the finance team might be privy to billing information that developers are not. To create this level of permissions, developer portals have to be architected in a more advanced manner.

While developer portals are focused on user-specific information, there is also a wide range of public information that is surfaced through developer documentation pages such as API status and changelogs, which you will learn about in the upcoming sections.

API status

An API status page is a simple page that displays the operational status of all the APIs for customers to reference. These statuses communicate key information to users about the state of the operation of the APIs at that particular time. For existing customers who may be experiencing errors or issues, this is a key piece of information in establishing whether the issues they are facing in their application are caused by an outage or degraded performance in your APIs.

The following screenshot shows how the Twitter API status page (`https://api.twitterstat.us/`) is designed to show various statuses of the APIs: operational, degraded performance, partial outage, major outage, and maintenance. This information is also followed by a log of all past incidents:

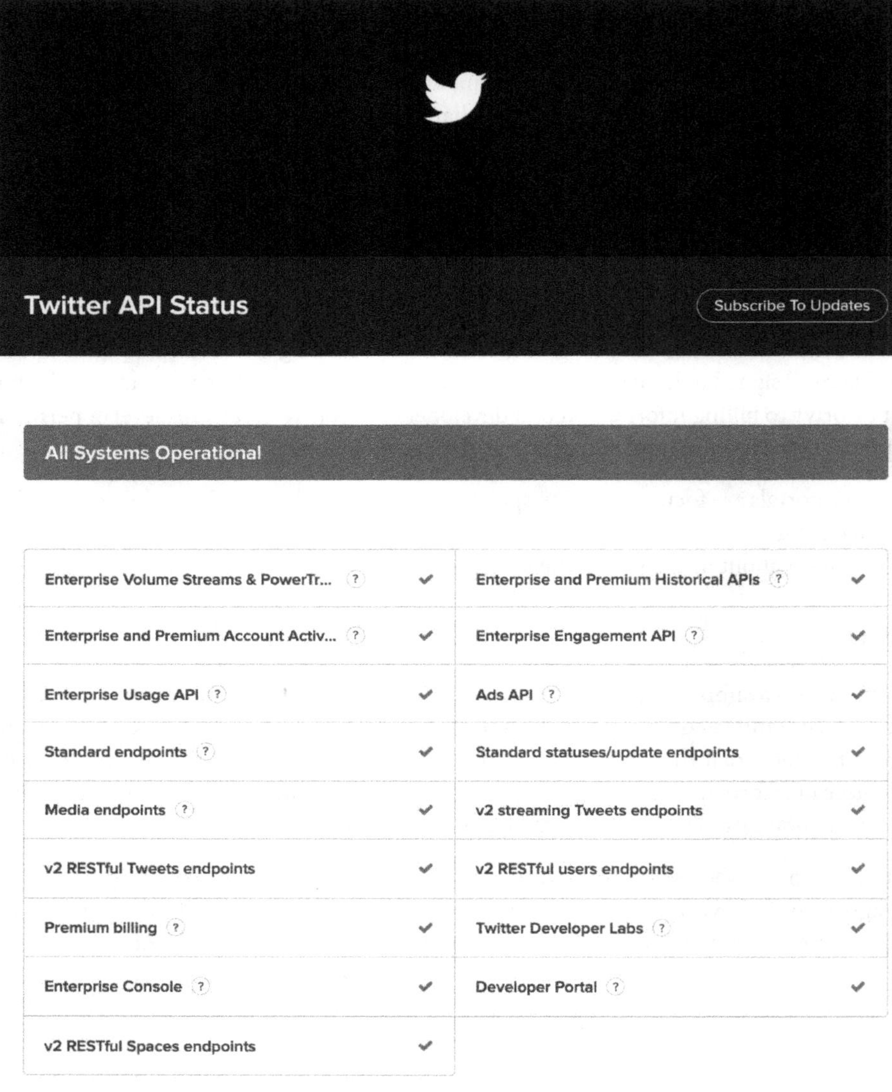

Figure 9.4 – Twitter's API status page showing the operational status of all systems

Oftentimes, DevOps engineers or on-call engineers on customer teams would be the team members looking for this information when they have incidents reported to them that they are trying to debug. If they are able to establish that your APIs are operational, they can continue their debugging process across other components of their application. An up-to-date API status page can save developers significant time in the event of an outage.

In the event of an incident, you should have alerts and notifications sent to the necessary people within your team to help resolve the incident as soon as possible. You should also instrument a mechanism to inform customers of outages as soon as possible to alert them. This would ensure that they are aware of any downstream impact on their application that your outage might cause. Setting up email notifications along with social media messaging for outage status messages is very useful.

API changelog

As your API evolves or you deprecate or retire APIs, you need a way to communicate the upcoming changes to customers, especially those that have existing integrations using these APIs. The API changelog page allows you to make public announcements to your customer base about any changes to your APIs. Changelogs are also a great way to drive awareness for new and awaited features among customers.

The following screenshot shows an example changelog from Meta's Facebook Messenger Platform:

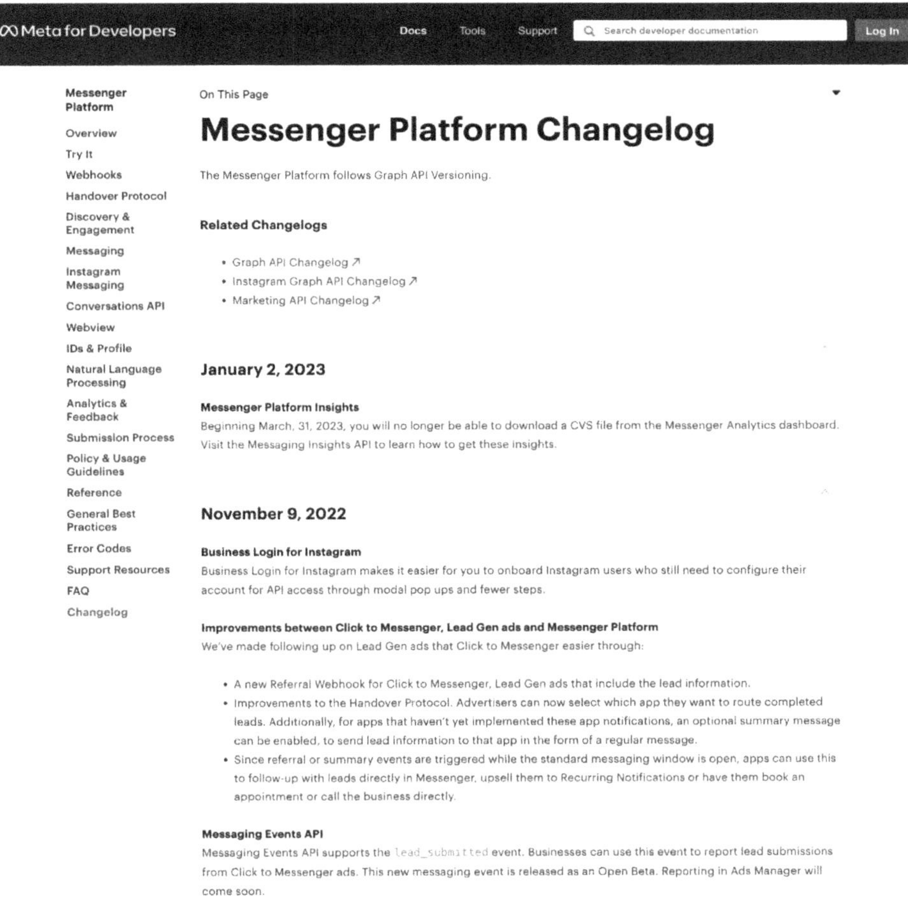

Figure 9.5 – Messenger Platform changelog

As you can see in the preceding screenshot, changelogs are structured like blog posts, but it is advisable to keep the messaging on the changelog as brief as possible. Changelogs are often confused with release notes, but changelogs are not just created for the purpose of documentation but also for customer communication. This means that changelogs are not only consumed by developers but could be viewed by any personas within the consumer base.

Changelogs are sorted by dates descending so that the latest changelog posts show up at the top of the changelog page. The language on changelogs must be easy to understand and must convey not only technical details but also the business impact of the changes.

In this section, you learned about the most fundamental component of API experiences, which is documentation. However, API experience can go beyond documentation to enable customers. You can use third-party tools as well as develop custom tools that help customers through various aspects of completing a successful integration.

Sample code and demos

Sample code and demos are resources that help developers see how the API can be used in practice and can serve as a starting point for their own projects. They are an important part of a good API experience, as they can help developers understand how to use the API and get started with their own projects.

Here are a few best practices for creating sample code and demos for an API:

- **Make them easy to understand**: Sample code and demos should be well documented and easy to understand, even for developers who are new to the API.
- **Include a variety of examples**: Providing a range of examples can help developers see how the API can be used in different contexts and scenarios.
- **Use common programming languages**: Using common programming languages can make it easier for developers to understand and use the sample code.
- **Keep the sample code up to date**: As the API evolves and changes, be sure to update the sample code to reflect any changes or new features.
- **Make them accessible**: Make sure the sample code and demos are easy to find and access. Consider hosting them on a developer portal or on the API documentation site.

By following these best practices, you can create sample code and demos that are helpful and useful for developers and that make it easier for them to get started with the API.

Integration guides and tutorials

Integration guides and tutorials are resources that provide step-by-step instructions for integrating the API into a particular environment or platform. They are an important part of a good API experience, as they help developers understand how to use the API and get started with their own projects.

Here are a few best practices for creating integration guides and tutorials for an API:

- **Make them easy to follow**: Integration guides and tutorials should be clear and concise, with step-by-step instructions that are easy to follow.
- **Include examples**: Including examples of how to use the API can be very helpful for developers, as it allows them to see how the API can be used in practice.
- **Provide troubleshooting tips**: It's inevitable that developers will run into issues while integrating the API. Providing troubleshooting tips and resources can help them resolve any issues they encounter.
- **Keep them up to date**: As the API evolves and changes, be sure to update the integration guides and tutorials to reflect any changes or new features.
- **Make them accessible**: Make sure the integration guides and tutorials are easy to find and access. Consider hosting them on a developer portal or on the API documentation site.

By following these best practices, you can create integration guides and tutorials that are helpful and useful for developers and that make it easier for them to get started with the API.

Providing developer tools

API documentation is an important part of the API experience because it provides developers with the information they need to use the API effectively. However, it is only one part of the puzzle. There are several other things that can help make the API experience better for developers, such as the following:

- **Sandbox**: A sandbox is a testing environment that allows developers to try out the API without affecting production data. This can be especially useful for testing out new features or debugging issues.
- **GitHub repositories**: If the API is open source, hosting the code on GitHub can make it easier for developers to contribute to the project and collaborate with the development team.
- **Developer communities**: Developer communities such as Stack Overflow provide a place for developers to ask questions, get help, and share their experiences with the API.
- **SDKs**: These are libraries that make it easier for developers to use the API, often by providing a higher-level interface that is more intuitive and easier to use.

- **CLI**: A CLI for an API is a way for developers to interact with the API using text commands rather than a **graphical user interface** (**GUI**).
- **Postman Collections**: Postman is a popular tool for testing and interacting with APIs, and Collections are pre-defined groups of API requests that can be saved and shared. Providing developers with Postman Collections can make it easier for them to get started with the API.
- **Low-code and no-code tools**: Low-code and no-code developer tools are platforms that allow developers to build applications or integrations without having to write a lot of code. Instead, they use visual interfaces and pre-built components to create the desired functionality.
- **Support resources**: These include things such as forums, FAQs, and email support, which can help developers troubleshoot problems they encounter while using the API.

These are just a few examples, but there are many other developer tools and resources that can be included in a good API experience. The key is to provide developers with the information and tools they need to effectively use the API and integrate it into their projects. In the subsections, you will learn more about these tools and how they contribute to a great API experience.

Sandbox

An API sandbox is a feature that allows you to simulate and test APIs. These activities are critical for developers. Before integrating APIs in the production environment, they must be aware of potential integration errors.

As a result, a proper environment that replicates the actual use of the API is required. The following screenshot shows the accounts page of PayPal's developer portal, where you can create and manage sandbox test accounts of two types—business and personal—to mock seller and buyer personas in a sandbox:

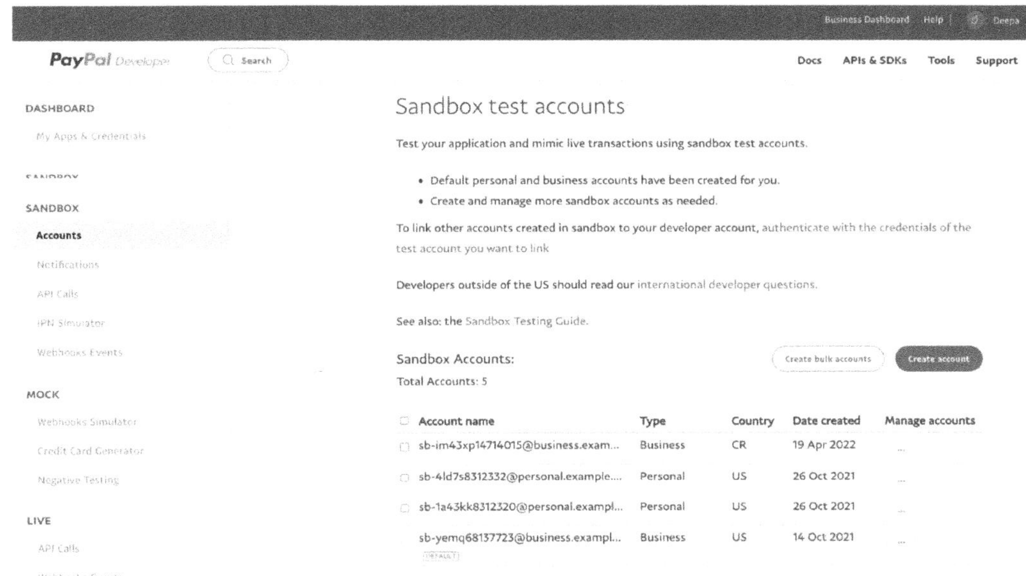

Figure 9.6 – Sandbox accounts view in the PayPal developer portal

The PayPal developer portal allows you to create business and personal accounts in a sandbox environment that is compatible with any APIs you want to test. You can make credit card transactions or PayPal transactions between these sandbox accounts and also use these accounts to log in to a UI-based sandbox experience to check the balances as they change. You can use the sandbox UI to test scenarios such as requesting refunds as well.

This extensive testing functionality allows developers to simulate production as closely as possible so that they can feel confident about their integration before going live. Such an advanced sandbox environment can be complex and expensive to develop, and now API producers are turning to less resource-intensive ways to enable testing, one of which is Postman Collections, which you will learn about in the next section.

Public GitHub repositories

Publishing a GitHub repository with use-case-based examples of code that your customers can reference or copy to build their applications is a great way of making the developers' jobs easier and reducing the time to integrate for your customers. This allows you to leverage the GitHub developer community of over 83 million developers who already use GitHub.

Publishing GitHub repositories will allow your customers to easily fork your repository and get started building applications quickly. You can track forks and get community feedback to engage with and learn from them.

Developer communities

Developer communities can be an important part of a good API experience because they provide a place for developers to ask questions, get help, and share their experiences with the API. There are several ways in which developer communities can contribute to a good API experience:

- **They provide a place to ask questions**: Developer communities can be a great resource for developers who have questions or are encountering issues while using the API. Other developers or members of the API team may be able to provide guidance or help troubleshoot problems.
- **They facilitate sharing of knowledge and experiences**: Developer communities can be a place for developers to share their experiences with the API, including any tips or best practices they have learned. This can help other developers get the most out of the API and avoid common pitfalls.
- **They help build a sense of community**: Developer communities can help create a sense of community and belonging among developers who are using the API. This can be especially important for developers who are working on their own or who are new to the API.
- **They can provide feedback**: Developer communities can be a valuable source of feedback for the API team, as developers may have suggestions for improving the API or ideas for new features.

In addition to major developer tools such as GitHub, which has a large developer community, Stack Overflow, Reddit, and dev.to are also emerging as great platforms to engage with the developer community. Developers go to Stack Overflow to ask and respond to questions from other developers who might be running into the same problems.

You can monitor keywords on these platforms to analyze the nature of the issues that your users are running into and also actively try to answer their queries and unblock them. Partner with developer relations teams to create an active practice of engaging with the developer community across platforms that developers use to get continuous feedback, and learn from the community of developers to make your APIs better.

SDKs

SDKs are libraries that make it easier for developers to use an API. They provide a higher-level interface that is more intuitive and easier to use than the raw API and often include additional functionality that can be useful for developers.

A way to think about SDKs and APIs is that APIs are like kits. If you are trying to build a table, you can either start with wood planks and work your way through building a table or, if you have a kit (such as one from Ikea), you have pre-cut wood pieces that are put into a kit to build a table.

Although they can be used on websites or other digital platforms, they are most typically utilized for mobile applications. They serve as a ready-made tool or application part. A well-designed SDK simplifies application development and improves the performance of applications by providing additional ways to reach and engage your users.

There are several ways in which SDKs can help improve the overall API experience:

- **They make it easier to use the API**: SDKs provide a more intuitive interface for interacting with the API, which can make it easier for developers to get started and integrate the API into their projects
- **They reduce the amount of code that needs to be written**: SDKs often provide a higher-level interface that requires less code to use, which can save developers time and effort
- **They provide additional functionality**: SDKs may include additional features or functionality that can be useful for developers, such as error handling, data parsing, and authentication
- **They simplify complex tasks**: SDKs can simplify complex tasks by providing a simpler interface for interacting with the API and handling common tasks

SDKs can help improve the API experience by making it easier for developers to use the API, reducing the amount of code that needs to be written, and providing additional functionality and simplification for complex tasks. Another such tool that is useful for developers is a CLI, which you will learn about in the next section.

CLIs

A CLI is a way for developers to interact with a computer or application using text commands rather than a GUI. In the context of an API, a CLI can be a useful tool for developers to perform tasks such as making API requests, managing resources, and interacting with the API in a terminal or Command Prompt.

The role of the CLI in a good API experience is to provide developers with an alternative way to interact with the API that is faster, more efficient, and more powerful than a GUI. For example, a CLI can allow developers to quickly perform tasks such as making API requests, managing resources, and interacting with the API without having to open a web browser or use a graphical interface. This can be especially useful for developers who prefer the command line or who want to automate tasks using scripts.

Some examples of tasks that a CLI for an API might support include the following:

- **Making API requests**: Developers can use the CLI to make API requests and view the responses, which can be useful for testing and debugging
- **Managing resources**: A CLI can allow developers to create, update, and delete resources using text commands
- **Interacting with the API**: A CLI can provide a range of commands for interacting with the API, such as retrieving information about resources or triggering specific actions

The role of the CLI in a good API experience is to provide developers with a fast, efficient, and powerful way to interact with the API and perform tasks using text commands.

Postman Collections

Postman has over 20 million users on its platform. Being an API tool, this means that all 20 million users are part of the target audience for any API product. For internal APIs, Postman Collections are a way for developers to collaborate and test APIs. In large organizations, internal Postman Collections also help with the discoverability of APIs across the organization.

The public workspaces in Postman are particularly useful for getting your APIs discovered by new audiences. Getting positioned on Postman's public network allows API producers to use the 20 million user base of Postman as a growth mechanism. The most prominent API producers, such as Meta, PayPal, Stripe, Twilio, and so on, publish official Postman Collections of their APIs to enable customers to easily get started with trying out their APIs in a familiar interface. Developers can start exploring APIs via Postman Collections without having to write any code, which also helps them get working examples of any requests they want to make in case they get errors.

The easy-to-use interface of Postman also allows enables low-code/no-code users who are either trying to generate some API code or doing business use-case testing using Postman. You will learn more about the tooling you can build to enable low-code and no-code users in the upcoming section.

Low-code and no-code tooling

The persona of the developer is expanding, and there is a new sub persona of the low-code/no-code developer that has emerged over time. These are developers who may not be coding but might be using tools to write code for them with little to no coding required. These customers are also building applications with your APIs and should not be ignored in your API experience. If your API experience is too complex for these users, or if you don't have any tooling to support their needs, you will miss the opportunity to address the needs of smaller business users and developer teams that do not have the resources of large enterprises and expert developers.

Low-code and no-code developer tools are platforms that allow developers to build applications or integrations without having to write a lot of code. Instead, they use visual interfaces and pre-built components to create the desired functionality.

Low-code tools require some coding knowledge but allow developers to build applications using a visual interface and pre-built components. This can reduce the amount of code that needs to be written and make it easier for developers to build applications quickly.

No-code tools are platforms that do not require any coding knowledge. They allow developers to build applications using a visual interface and pre-built components, without having to write any code. No-code tools are often targeted at business users or non-technical individuals who want to build custom applications or integrations.

Both low-code and no-code tools can be useful for building integrations and applications quickly and can be a valuable addition to a good API experience. They can make it easier for developers to get up and running with the API, even if they don't have a lot of coding experience.

Twilio recognized the need to enable low-code and no-code developers and launched Twilio Studio (`https://www.twilio.com/serverless/studio`), which provides a drag and drop GUI for customers to build applications using Twilio APIs. The following screenshot shows the UI of Twilio Studio with a canvas where users can design call flows using pre-built widgets from a widget library shown on the right:

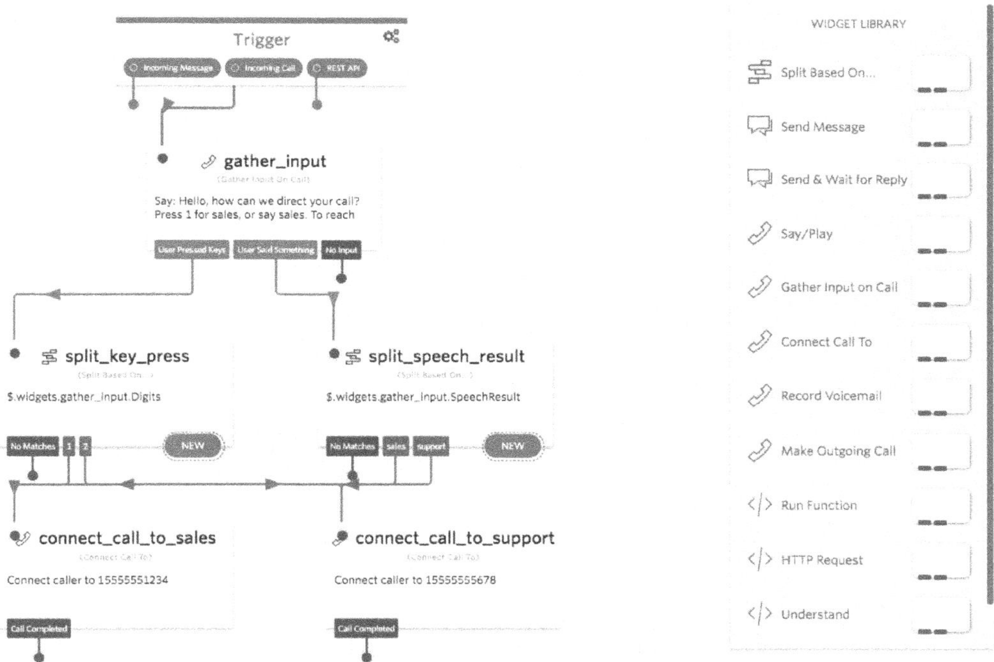

Figure 9.7 – Twilio Studio's drag-and-drop interface to work with APIs

Users can use this drag and drop interface to create, edit, and manage voice, messaging, and communications apps. This lowers the barrier to entry for users to be able to use Twilio APIs and build SMS surveys, voice applications, notifications, SMS chatbots, and so on without the need to write any code.

Stripe has created a click-through experience (`https://checkout.stripe.dev/`) for the standard use cases for payment APIs. In the following screenshot, you can see Stripe's Checkout click-through builder page 1, where the user can configure brand colors, style, and font, and also enable or disable functionality such as coupons, shipping, tax, and so on:

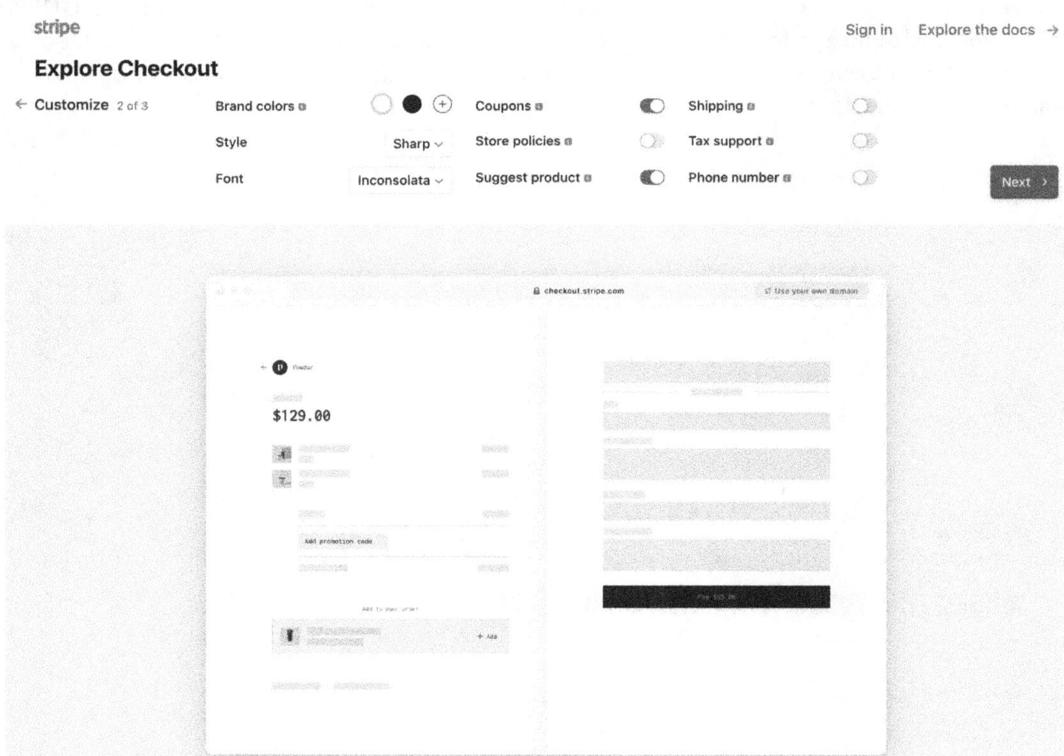

Figure 9.8 – Stripe's Checkout click-through builder (page 1)

Once the user has made the selections on page 1, they can move to the next page of the Checkout builder where they can select from Apple Pay, Google Pay, or other payment options and view the Checkout update with the selections:

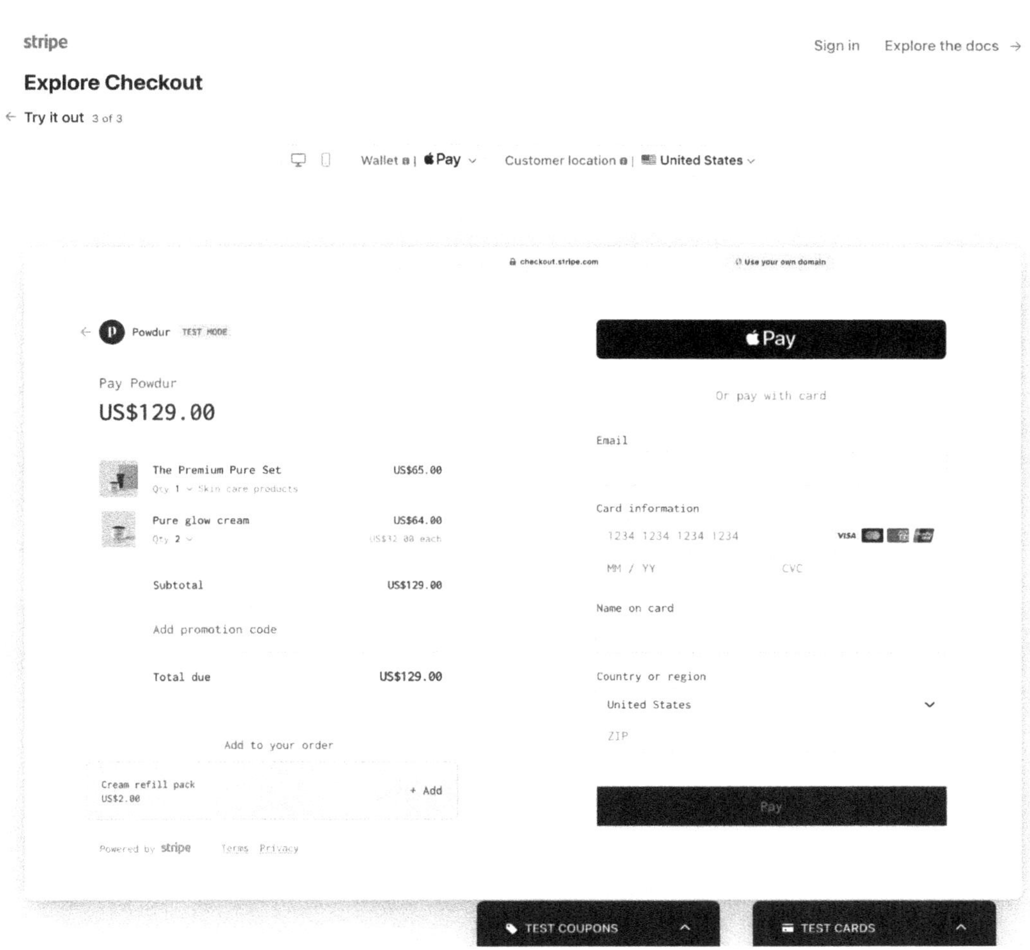

Figure 9.9 – Stripe's Checkout click-through builder (page 2)

This interface is also equipped with test coupons and test cards that are handy for customers to be able to experience the checkout flow as they design it. Once the customer is satisfied with the configuration, they can copy and paste the code into their website and complete their integration without having written any code. This also actively reduces the overall time to integrate for your customers who are only looking for standard use cases and reduces errors they might have run into in the process.

Building low-code/no-code tooling can be rewarding based on the nature of your business and the target audience you are trying to address.

Instrumenting support mechanisms

As you learned in *Chapter 6, Support Models for API Products*, support is a crucial aspect of customer experience. Making sure customers can find support channels when they need them can take many forms across various components of the API experience.

Live chat support is a great way to let customers reach out for help as they go through the products and evaluate them, but this can often be resource intensive. Products such as Intercom have innovated significantly in creating messaging experiences that find the balance between resource-intensive support and self-help for customers.

Oftentimes, developers will have questions about your APIs that are not clearly addressed in the documentation, or if they run into complicated scenarios during the integration process, they might need support. Tools such as Zendesk enable you to build support flows while allowing customers to choose their preferred channel of communication.

Direct support channels such as Zendesk tickets and chat don't always capture the platforms that your customers are on. Social media channels and developer communities often have developers posting questions when they get stuck. Developer relations teams frequently monitor and respond to community platforms, where they can also identify trends in customer pain points and report them to product managers for improvement, as well as create tutorials to explain better.

Although there is a large developer community online, organizing workshops and events is still a great way to leverage the developer community around your APIs and get a continuous flow of feedback. Enabling developers to be aware of your APIs allows them to be able to guide and unblock other developers, both in their organizations and in online communities.

Hosting or sponsoring conferences is another way many API companies raise awareness and build a community around their APIs. To create a great API experience, you need to think about how to document it well and give your customers the tools they need to integrate with your APIs quickly and easily.

Summary

In this chapter, you learned about the components of API experiences across documentation and developer tools that you can consider offering to your customers. In the early phases of launching your APIs, it's possible that you only provide a subset of this vast set of tools, but over time, you can learn about the needs of your customer base to improve and expand your API experiences.

The starting point of API experience for any API is a simple API reference. Once you begin with API references in the beta phase, you can learn from your customers and invest in building more tools to enhance the experience. As your API matures, you can invest in tools that make the API experience richer.

A great API experience with good documentation, developer tools, a sandbox, Postman Collections, support resources, SDKs, and so on can help customers easily understand how to use the API and get started with their own projects, saving them time and effort. Overall, a great API experience can make it easier and more efficient for customers to use the API, which can help them be more successful in their projects and achieve their goals.

Now that you have learned about the various components of the API experience and the role of the product manager in the success of your APIs, you will learn about setting metrics across infrastructure, product, and business for your APIs in the following chapters.

Part 3: Deep Dive into Key Metrics for API Products

The goal of any product is to drive value for its customers. If customers can discover your product, easily understand the utility it provides, and start using it effectively in a way that they find valuable, your product has successfully achieved its goal.

In *Part 1*, you learned about the roles and responsibilities of an API product manager and how to build, grow, maintain, and support APIs as products. In *Part 2*, you learned about the developer journey and how the consumers of API products discover, evaluate, integrate, test, and measure the APIs they consume. Now that you have a deep understanding of the customer, you can start to think about metrics that can measure various aspects of a product across producers and consumers. You will learn to manage APIs methodically and learn about the levers you have to improve these metrics so that your product can be successful.

Translating touchpoints to measurable metrics

Using steps of the customer journey to establish metrics to measure a product's performance is a useful approach because it allows you to understand how customers interact with your product and identify areas for improvement. By identifying and tracking metrics that correspond to each stage of the customer journey, you can gain a deeper understanding of how customers interact with your product and identify areas for improvement. This can help to improve the overall customer experience and drive growth for the business.

In *Part 2*, we broke down each interaction the customer has with a product into individual touchpoints; in *Part 3*, you will learn to measure each of these steps and identify patterns in user behaviors that work in favor of or against your product goals. By following the developer journey, you can identify the different steps customers take to successfully use your product. In this process, you will also be able to identify where your customers struggle. In *Part 3*, you will learn to translate the steps of the developer journey into a measurable conversion funnel and the methods to identify signals, such as activation, engagement, retention, and scale.

Across the user journey, users go through discovering a product, learning and adopting the product, and then either scaling or declining their usage. All of the components of the API experience drive the discovery and adoption of API products.

A holistic approach to analytics

There are three core dimensions to API product analytics that measure a product across all aspects of the customer journey – infrastructure, product, and business, as shown in the following diagram.

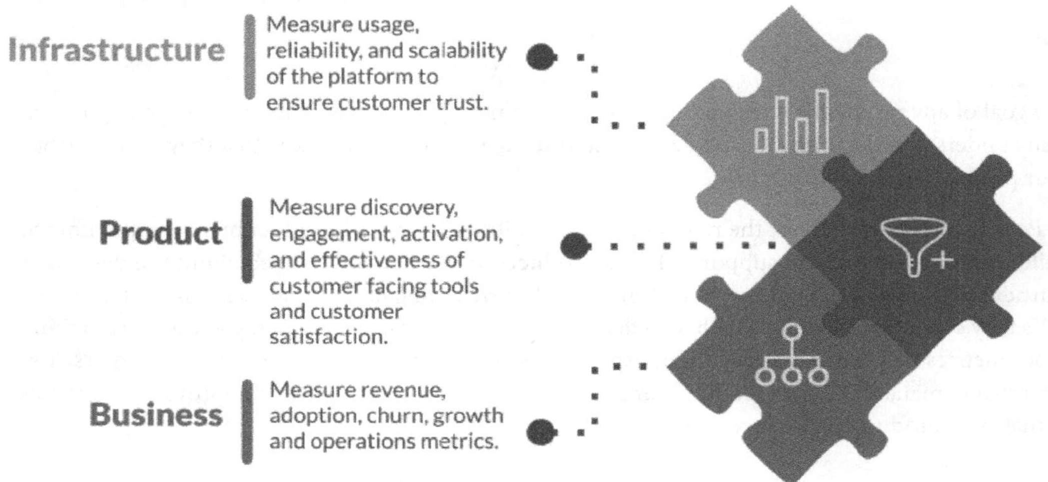

Infrastructure
Measure usage, reliability, and scalability of the platform to ensure customer trust.
Product
Measure discovery, engagement, activation, and effectiveness of customer facing tools and customer satisfaction.
Business
Measure revenue, adoption, churn, growth and operations metrics.

Figure P3.1 – Dimensions of API Analytics

Being data-driven in API management is important because it allows companies to make informed decisions about their API strategy, operations, and performance. A data-driven approach enables a company to understand how its API is being used, how it's performing, and how it's impacting a business.

Here are some examples of why a data-driven approach is important in API management:

Infrastructure: API infrastructure metrics such as uptime, response time, and error rates can be collected and analyzed to identify and troubleshoot issues, and to ensure that an API is performing well and meeting service-level agreements.

Product experience: Metrics such as user engagement, user satisfaction, and conversion rates can be collected and analyzed to understand how well an API is meeting customer needs and how it is impacting a business.

Business: Business-related metrics such as revenue, customer acquisition, and retention can be collected and analyzed to understand how an API is impacting the overall performance of a business.

By being data-driven, companies can make informed decisions about their API strategy, operations, and performance. For example, by analyzing the usage data, companies can understand which features are being used most and which are not, and make decisions on which features to invest more in. Additionally, by analyzing performance data, companies can identify bottlenecks and make decisions on scaling the infrastructure accordingly.

Overall, being data-driven in API management allows companies to make informed decisions about their API strategy, operations, and performance by providing a clear picture of how an API is being used, how it's performing, and how it's impacting a business.

In *Part 1*, you learned about how APIs are built and the growth goals that drive the development of API products. In *Part 2*, you learned about how API products are discovered, evaluated, and integrated by developers on the customer side and how you can better understand the touchpoints to help API customers be successful. In *Part 3*, you will learn about how to establish specific metrics that tie together the goals of the API producers and API consumers so that you can measure whether those goals are being met.

In the chapters of *Part 3*, you will learn about the language, and how to calculate and understand the most important metrics in all areas of APIs. The following chapters will be covered in this part:

- *Chapter 10, Infrastructure Metrics*
- *Chapter 11, API Product Metrics*
- *Chapter 12, Business Metrics*

10
Infrastructure Metrics

API producers build APIs for a number of customers, both internal and external. From a technical point of view, any API must meet certain engineering standards for your software architects to accept it. As we discussed in *Chapter 3*, engineering standards control how the APIs are built, tested, secured, and put into use. Once your APIs are in production, however, it is important to start monitoring their performance and usage.

In this chapter, you will learn how reliability is defined and how the reliability of APIs delivers value to customers. The topics we will be covering in this chapter include the following:

- **Key success factors** (**KSFs**) for APIs
- Infrastructure as the foundation for API analytics
- Performance metrics
- Usage metrics
- Reliability metrics

The first step in giving your users a high-quality experience is to have well-defined metrics for how your API works and how it is used.

Key success factors (KSFs) for APIs

KSFs are critical areas or elements that are necessary for the success of an API. Identifying the KSFs for an API can help you understand what you need to focus on to ensure your APIs are successful.

For an API, the KSFs can vary depending on the specific use case, but some examples could include the following:

- **Performance**: The API should have a low response time, high throughput, and a low error rate
- **Scalability**: The API should be able to handle a high number of requests and increases in traffic
- **Reliability**: The API should be available and responsive at all times, with minimal downtime

- **Security**: The API should be secure and protect sensitive data
- **Usability**: The API should be easy to use and understand for developers
- **Flexibility**: The API should be flexible and able to integrate with different systems and platforms
- **Adoption**: The API should be widely adopted by developers and organizations
- **Retention**: The API should be able to retain its users and create a loyal user base
- **Revenue**: The API should generate revenue for the organization, either through monetization or by creating value for it
- **Innovation**: The API should be able to create new opportunities and bring innovation to the industry

By focusing on these key areas, an organization can increase the chances of success for its API and ensure it is providing value to both the organization and its users.

In order to ensure that you are evaluating and monitoring your APIs for these factors, you need an analytics strategy that addresses all these factors. Across *Chapters 10*, *11*, and *12*, you will learn about a three-part framework for establishing metrics for APIs: infrastructure, product, and business. Infrastructure metrics attempt to measure the technical aspects of the APIs, such as performance, usage, and reliability. Product metrics include the measurement of all aspects of the user journey and user experience, while business metrics measure business processes and outcomes such as operations and revenue. In the following section, you will learn about how infrastructure sets the foundation of API analytics for APIs.

Infrastructure as the foundation for API analytics

If you build a website and it takes too long to load, it does not matter how amazing its design is because a good portion of the customers will bounce off the web page without experiencing it. If you create a checkout page to accept payments and the payments only go through 80% of the time while erroring out the other 20%, you will be losing 20% of the revenue. These are performance-related issues that determine the user experience without any dependency on the design of the product. In the case of APIs, we tend to spend a lot of time thinking about design and developer experience, but a key aspect of the API experience is the performance of the APIs, which would determine their reliability and scalability, and the customer's ability to use the APIs.

In the same way that a customer doesn't consciously measure the page load times of a web page but only responds to the slowness in a negative way, customers may not be able to measure the infrastructure metrics of your APIs, but this will significantly impact their experience when using the APIs. Your customers, across internal, partner, and public APIs expect the highest level of quality and reliability from your APIs. This quality inspires them to integrate your APIs into their applications. The infrastructure of your APIs is a black box to your customers. They don't know how your APIs are implemented, which tools you use, or which gateways you use. The customer only cares that the APIs work and that they are responsive, reliable, and scalable. To enable this black-box experience, your engineering teams will have to work on various aspects of the API infrastructure, and oftentimes as your user base scales, scaling your infrastructure will become a priority for your product roadmap.

Infrastructure metrics form the foundation of metrics that you establish for your API product. This is because if the APIs aren't performing at an acceptable level, the work you do to enable adoption will be wasted and, consequently, revenue goals will not be met. It's only when you are able to deliver stable and reliable APIs that you can expect users to trust your APIs and build their applications using them. *Figure 10.1* shows how infrastructure, product, and business metrics are building blocks, with infrastructure being the foundational building block and product and business built on top of it:

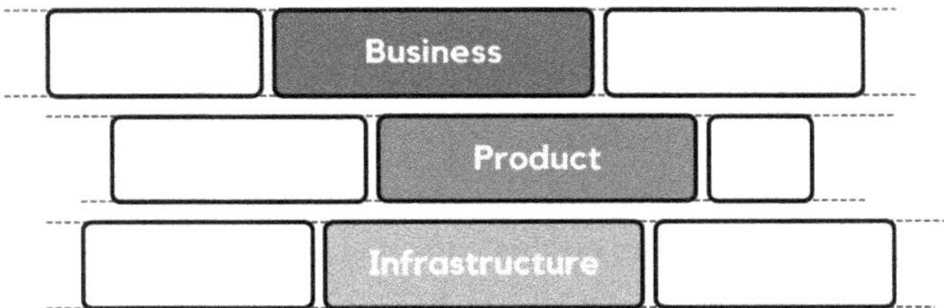

Figure 10.1 – Infrastructure as the foundational building block for API analytics, which enables product metrics and business metrics to be built on top of it

As the producer of the APIs, you should know how well the APIs work. If you don't monitor to see how reliable your APIs are, you'll have to hope that everything works as planned. If your API breaks, the people who use it—internal collaborators, partners, or other developers from outside your company—will find out and report it to you. Continuous monitoring can help you find these problems before they worsen and impact your customers.

The developers who use your API are dependent on how well it works. To establish performance and reliability-based **service-level agreements** (**SLAs**), you must monitor your APIs closely at all times.

In addition to business value, reliability is an important aspect of decision-making for your customers regarding whether they will use your APIs or not. If your APIs were reliable, that would inspire trust in your customers. But what does *reliability* mean? A reliable API has the following qualities:

- **Consistent**: A set of APIs can be considered consistent if they are documented properly and are predictable in their design and operation.
- **High availability and uptime**: APIs that have significantly high levels of uptime and do not experience frequent downtime can be considered reliable. The acceptable level of uptime and availability can be defined as part of the SLAs in your terms and conditions.

- **Latency**: Latency refers to whether the speed at which an API responds to requests is slow or fast.
- **Secure**: An API should have well-documented terms and conditions in terms of its usage, along with control over who has access to it.
- **Status**: Any outages, planned maintenance, and changes should be documented and exposed to users so that they can plan to manage their applications and customers who will be impacted by them.

The true benefit of monitoring is that it recognizes patterns and outliers, whether you run it once or regularly. Realizing how your API performs over time is crucial. Continuous API monitoring allows you to do the following:

- **Examine patterns across time**: At what rate do requests increase?
- **Check the progress of an API's performance over time**: Have you noticed a slowdown in the response time during the past week?
- **Debug your API**: It is essential you do this. In scenarios where error rates recently rose for your API, you should look for what else has changed simultaneously.

Incorporating testing into your monitoring strategy is a smart move. Make sure your testing spans authentication, availability, and performance. Test and monitor your APIs frequently since many things can impact your APIs' performance, and you'll be able to pinpoint precisely when performance issues start cropping up.

For transparency with your API users, you can compile alarms, test results, and other data into a status page. Building confidence requires a status page that displays which endpoints are operational and a history of outages or problems.

Getting your API monitoring set up early is always better than starting late. APIs should be monitored and tested during their whole life cycle. This API monitoring will provide your teams with the information they need to make your API more stable. In the following sections, you will look at the three major categories of infrastructure metrics—**performance**, **usage**, and **reliability**—and the individual metrics within these categories, as shown in the following diagram:

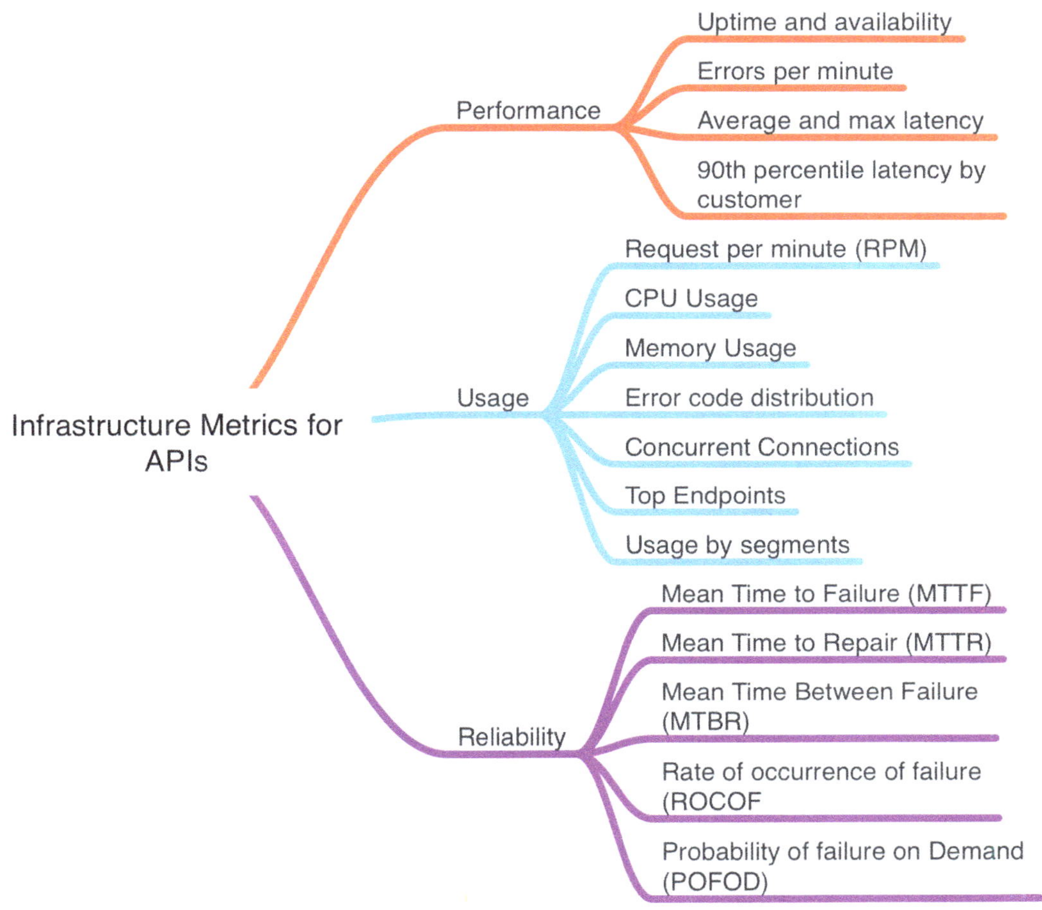

Figure 10.2 – Infrastructure metrics across performance, usage, and reliability

Depending on the various tools you use, you will uncover various aspects of strengths and weaknesses in your APIs using this set of metrics. The insights from these metrics should drive the priority of engineering efforts as well as your product roadmap to improve the quality and reliability of your APIs.

Performance metrics

For APIs, as for any other product, you can measure the performance of the product from the most fundamental perspective to determine whether the product works the way it is expected to. Although it might seem like an abstract quality to measure, there are ways you can quantify the performance of APIs using the metrics shown in the following screenshot:

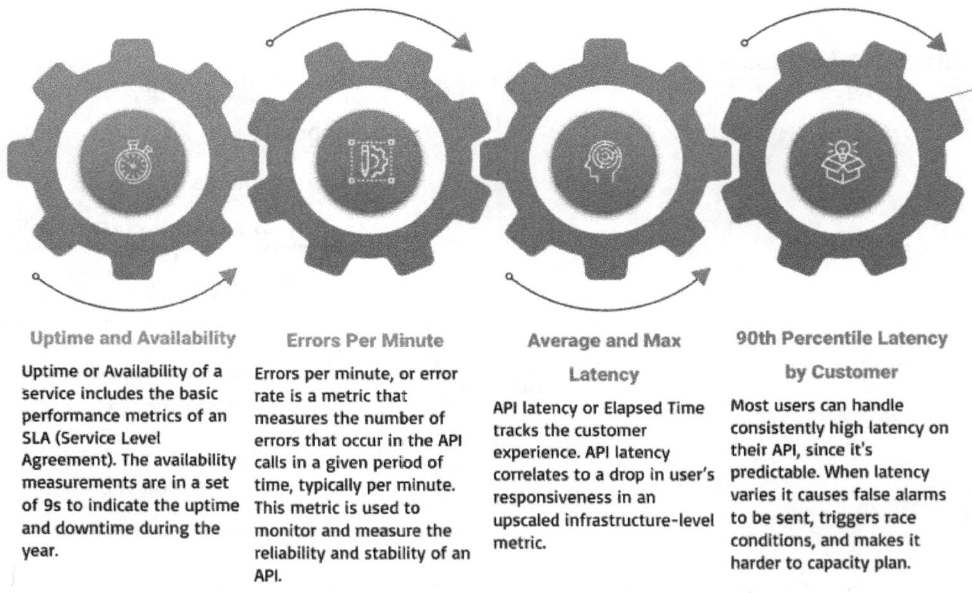

Uptime and Availability
Uptime or Availability of a service includes the basic performance metrics of an SLA (Service Level Agreement). The availability measurements are in a set of 9s to indicate the uptime and downtime during the year.
Errors Per Minute
Errors per minute, or error rate is a metric that measures the number of errors that occur in the API calls in a given period of time, typically per minute. This metric is used to monitor and measure the reliability and stability of an API.
Average and Max Latency
API latency or Elapsed Time tracks the customer experience. API latency correlates to a drop in user's responsiveness in an upscaled infrastructure-level metric.
90th Percentile Latency by Customer
Most users can handle consistently high latency on their API, since it's predictable. When latency varies it causes false alarms to be sent, triggers race conditions, and makes it harder to capacity plan.

Figure 10.3 – Infrastructure metrics to measure the performance of APIs

As you can see in the preceding screenshot, the performance of your APIs can be measured using four metrics, which we will discuss in the following subsections.

Uptime and availability

Although customer applications depend on the performance of your APIs, the choices and nuances of your infrastructure choices that make your APIs performant are abstracted from your customers. Uptime and availability are the most outward-facing of all infrastructure metrics. Most companies would publish a status page with uptimes and availability stats of their APIs. These pages are generated using a ping service.

The availability of a service is the probability of the system being available to the user over a period of time. Since availability is the likelihood of the system being available, it helps you understand the reliability of your APIs at a wider scale and how it might impact the user experience.

Uptime measures the reliability of an API as a percentage of time the service has been working and is ready for use. Uptime is used as the go-to standard for measuring the availability of APIs. In the following screenshot, you can see how Stripe displays the uptime of its APIs over a 90-day period on its status page (`https://status.stripe.com/`):

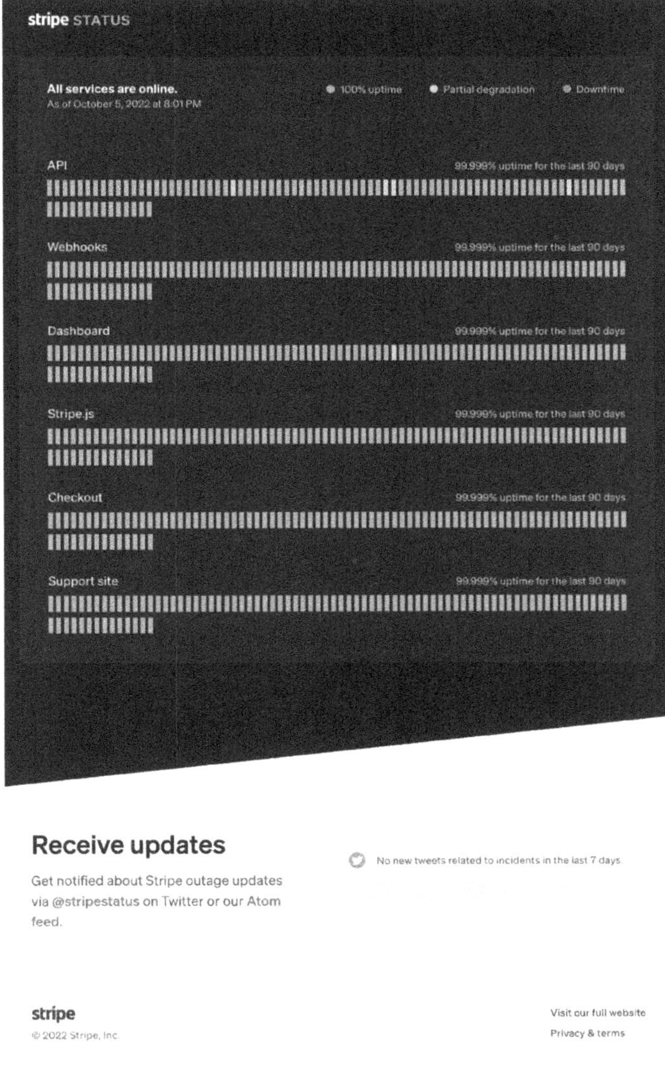

Figure 10.4 – Stripe API status page displaying uptime metrics over a 90-day period

In the preceding screenshot, you can see that the uptime stats of Stripe's APIs also display partial degradations as yellow bars in the middle of mostly green bars representing 100% uptime. When new customers review Stripe APIs as a possible solution, this level of transparency in reporting uptime inspires confidence that the service is reliable. For customers who already use Stripe, this page can be a very useful source of information in case of incidents with their own applications. They can quickly access this page to determine whether the incidents they are facing could have a dependency on Stripe APIs and rule out the possibility of this being the case.

Uptime provides insight into the past performance of a service but is not a guarantee of performance in the future. This is important to understand since outages can still happen, and uptime is merely a reflection of the trends of current performance. Needless to say, if a service is consistently experiencing outages or if the uptimes are not suitable for customer expectations, the customer can decide against using your APIs. This is why SLAs on enterprise agreements often include specific performance metrics such as uptime—usually in a set of 9s, such as 99.9999% availability. 100% availability is considered a myth, and an almost 100% uptime of 99.99% is often considered realistic and ensures a good customer experience.

Five Nines is the term often used to describe 99.999% when it comes to uptime and availability. An availability level of **1 None** represents 90% uptime. 90% uptime equates to your service being down for 36.5 days a year. Availability is directly proportional to uptime, and any increase in availability will be accompanied by an increase in uptime. 99.999% of uptime is 5.26 minutes of downtime a year. To put this in context, Google experienced a 25-55-minute outage in 2014 that impacted applications such as Gmail and Google+, affecting approximately 10% of the user base. There have been several outages across Facebook, Salesforce, Twilio, and so on over the years, impacting millions of users. 100% uptime is impossible and downtime is unavoidable, but it continues to drive impact and customer experience.

Understanding both uptime and availability is important because incorrect assumptions can result in an SLA breach and end up costing you. These metrics must be reviewed accurately to ensure you meet the thresholds set by the agreements because by not doing so, you will be delivering a poor customer experience.

Errors per minute

A non-`200` response for an API constitutes an error response. API error responses that range in the 4XX and 5XX categories constitute errors. Error responses in the range of 4XX may indicate multiple issues not limited to infrastructure issues. To measure how buggy or error-prone your APIs are, you can measure the rate of error in terms of **errors per minute** (**EPM**).

When your customers are using your APIs and building their applications using them, unexpected errors will impact their customer experience. If your APIs are prone to errors, this will also increase the development effort required to integrate with your APIs, in addition to damaging the trust your customers have in your service.

User system errors can result from poorly designed APIs and can negatively impact the experience for your customers. Tracking the EPM can also be insightful in detecting threats. For example, an increase in unauthorized 401 errors from a specific geographical region might be an indication of system bots trying to hack your APIs.

Average and maximum latency

The latency of an API is the time it takes for the infrastructure to process and respond to an API request. This is the time it takes for the client to send its request to the server and receive the first byte of the response. This is also known as **time to first byte** (**TTFB**). The shorter the response time of your APIs, the better your user experience is going to be. The response rate is not to be confused with latency, since the response rate includes latency along with the time for the request to be fulfilled.

Average latency and maximum latency provide insight into how your APIs are changing over time. Track these metrics across API versions and schema changes as your API evolves to identify changes that lead to changes in responsiveness.

If the server does not have enough capacity to fulfill the number of requests coming to it at any given time, the responsiveness of your APIs will be impacted. When there is a bottleneck of requests, the server is otherwise overburdened, or requests are managed inefficiently, API latency rates might increase. To instrument monitoring for latency, you can use a ping service that can provide an insightful assessment of how the user experience is being impacted. To improve the latency of your APIs, invest in server speed and capacity to match the scale of your operation. A common practice to manage latency is to geographically distribute deployments to reach servers if your customers are worldwide.

90th-percentile latency by customer

90th-percentile latency is a core engineering metric for APIs. Since low averages can easily mask large variations in your latency, you can measure the 90th-percentile latency over the average. This allows you to remove the impact of spikes that averages can mask, which are more damaging to the customer experience than low latency. API monitoring tools such as Moesif allow you to monitor 90th-percentile latency broken down by endpoint.

Consistently high latency on an API is tolerable to the vast majority of users because it is predictable. False alarms are sent, race conditions are triggered, and capacity planning is made more difficult when there is unpredictable variation in latency. To get a better assessment of the performance of your APIs, you can target the 90th percentile instead of tracking the arithmetic mean.

High-performance APIs enable your users to reliably use your APIs as part of their applications. In the following section, you will learn about usage metrics you can track.

Usage metrics

You will hear the term *usage* repeatedly when it comes to metrics. This is because there are various dimensions of usage metrics. In terms of infrastructure, you are measuring the usage of the resources of your API infrastructure. When we talk about usage as part of product metrics in *Chapter 11*, we will dive into the consumer experience and usage in terms of the product.

In the following screenshot, you can see the seven key infrastructure metrics that you must track across all your APIs:

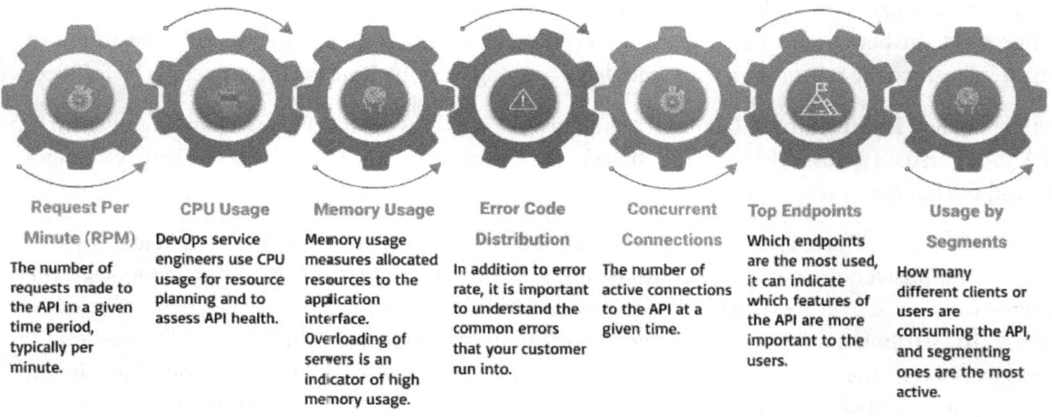

Request Per Minute (RPM)
The number of requests made to the API in a given time period, typically per minute.
CPU Usage
DevOps service engineers use CPU usage for resource planning and to assess API health.
Memory Usage
Memory usage measures allocated resources to the application interface. Overloading of servers is an indicator of high memory usage.
Error Code Distribution
In addition to error rate, it is important to understand the common errors that your customer run into.
Concurrent Connections
The number of active connections to the API at a given time.
Top Endpoints
Which endpoints are the most used, it can indicate which features of the API are more important to the users.
Usage by Segments
How many different clients or users are consuming the API, and segmenting ones are the most active.

Figure 10.5 – Key API infrastructure usage metrics

In the following sections, you will learn in-depth about these seven key infrastructure usage metrics.

Requests Per Minute (RPM)

The RPM metric is a standard metric for gauging the application manager's performance. Since the server doesn't account for the latency caused by **input and output** (**I/O**) processes, it acts as a ceiling for an application's interface.

RPM is commonly used to compare HTTP or database servers, as it is a metric of throughput. While some may take pride in a high number of RPM, it should be efficiency, not showmanship, that guides an engineering team's efforts. While a high number of RPM might sound like a good thing, efficiency should drive an engineering team's priorities.

To improve the RPM, it is necessary to design strategies that reduce the number of API calls made. You can reduce the number of API calls needed to perform certain business transactions to reduce the RPM. Having a flexible pagination mechanism and using common patterns such as batching many queries into one request may prove helpful.

If your API is designed for other businesses, you may notice that API traffic is lower on evenings and weekends, which means that your RPM reliance may change during different days of the week and times of the day.

CPU usage

CPU usage is an age-old indicator of performance that has been used for decades to manage resources and track patient wellness (a proxy for application responsiveness). High values mean the server/ **virtual machine** (**VM**) is oversubscribed/overloaded or the application has a bug. When the value is low, fewer servers need to be maintained or more programs can be active.

The amount of CPU time used is a crucial indicator of how quickly an API responds. The CPU utilization is used by DevOps service engineers for the purpose of resource planning and API health monitoring. **Machine learning** (**ML**) workloads and application encoding via floating-point differences are the direct result of the high server CPU usage that applications such as API gateways or bandwidth services inevitably experience.

In order to track metrics such as thread profiling, memory, and CPU usage, cryptographic network protocols can be embedded in to your application and monitored via system and programming monitoring. There may be improper threading in play if usage is skewed in one direction.

Memory usage

The utilization of both memory and processing power is intertwined. CPU utilization is a suitable proxy for assessing resource consumption since both CPUs and memory capacity are physical resources, unlike a measure, which may be more configuration-dependent. VMs with exceptionally low memory usage can be reduced, or their memory consumption can be increased by allocating more services to the VM. The converse is also true; high memory utilization may point to overburdened servers.

In the past, memory was always more of an issue than computing power when it came to production databases and big data queries/stream processing. Because additional RAM available can lessen checkpointing, network synchronization, and paging to disk, it is a strong indicator of how long your batch query can take. The number of page faults and I/O operations are additional important metrics to consider when analyzing memory use. An easily made error is an application set to allocate no more than a small percentage of available physical memory, which can lead to excessive page virtual memory thrashing.

Application performance monitoring (**APM**) solutions, such as Datadog, are the ideal approach to keeping tabs on these indicators. Infrastructure engineers occasionally notice that the production databases they are responsible for are using more memory than CPU. Memory use, however, can be used as a proxy for checkpointing, network synchronization, or the tally of page vaults and I/O operations.

Error code distribution

We looked at error rate or the EPM as part of performance metrics, but when it comes to usage, you should also monitor the different error codes your users are receiving. Among the various error codes, some might be expected while some may not. API analytics products such as Moesif provide a great way to view error code distribution across your product usage metrics, as shown in the following screenshot:

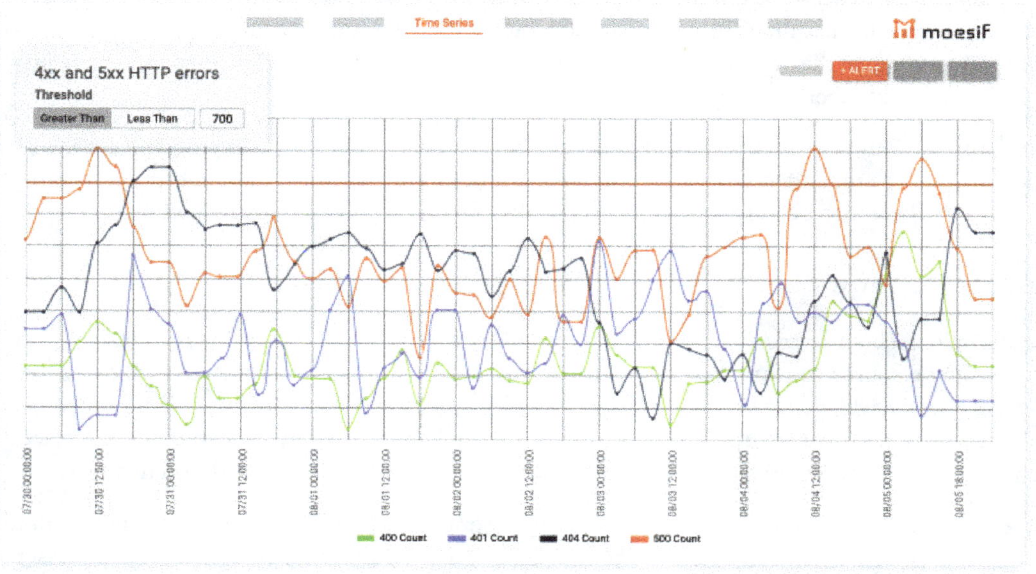

Figure 10.6 – Moesif dashboard showing the frequency of occurrence of 4xx and 5xx error codes

This Moesif dashboard helps you identify the patterns of errors in the 400 range (4xx) and 500 range (5xx). You can observe patterns over time to identify acceptable ranges and also spot spikes that might impact your customer integrations.

Concurrent connections

The concurrent connections metric for an API measures the number of active connections that are currently being used to communicate with the API at a given point in time.

This metric is important to monitor because it can indicate how many clients or users are currently interacting with the API and the level of traffic the API is handling. High concurrent connections can indicate that the API is being heavily used and may indicate a need for scaling or optimization of the API.

Concurrent connections can be measured in several ways, such as by tracking the number of open TCP/IP connections or by tracking the number of active sessions on the server side.

It's important to note that having a large number of concurrent connections does not necessarily mean that the API is not performing well, but it's an indication that the API is being heavily used and it's important to keep an eye on it.

Monitoring the concurrent connections metric over time can help identify patterns and trends in usage, and can be used to identify potential issues or areas for improvement. It can also help to set limits on the number of concurrent connections, to avoid overloading the API or the underlying infrastructure.

Top endpoints

The top endpoints metric for an API measures the usage of the various endpoints, or functions, of the API. It can help to identify which endpoints are being used the most and which ones are the most important to the users.

This metric can be collected by tracking the number of requests made to each endpoint, as well as the response time for each endpoint. This data can be analyzed to identify which endpoints are the most frequently used, which ones are the most resource-intensive, and which ones are performing poorly.

Knowing which endpoints are the most used can give insight into which features of the API are the most popular and which ones are the most valuable to the users. This can help with API development strategies, such as focusing on improving the performance of the most used endpoints, or the development of new features that are in high demand.

Also, it can help to identify which endpoints are not being used as much and consider deprecating or removing them to reduce the complexity of the API and improve its performance. When endpoints are not being actively used, they can also be a security risk as hackers often target such API vulnerabilities for cyber attacks.

The top endpoints metric can be monitored over time to track usage trends and help identify potential issues and reduce related infrastructure maintenance costs. It can be used in combination with other metrics such as the RPM, error rate, and response time to have a better understanding of the performance of the API.

Usage by segments

There are several types of user segmentation metrics that can be used for an API to understand and analyze the usage patterns of different groups of users:

- **Demographics**: This includes metrics such as age, gender, location, and the income of the users. This can help to understand the user base and target the marketing efforts.
- **Behavioral**: This includes metrics such as the frequency of use, the number of requests made, the number of errors, and the types of requests made by the user. This can help to identify the most active users and understand their usage patterns.
- **Device**: This includes metrics such as the type of device used to access the API, the operating system, and the browser. This can help to understand the user's environment and identify potential compatibility issues.
- **Cohort analysis**: This includes metrics such as the number of new users, the number of returning users, and the number of churned users. This can help to understand user retention and identify patterns of user engagement.
- **Funnel analysis**: This includes metrics such as the number of users that complete a specific task or reach a specific endpoint in the API. This can help to understand user engagement and identify potential areas of friction in the user experience.
- **Geolocation**: This includes metrics such as the geographic location of the users. This can help to understand the user base and target marketing efforts.

By segmenting users and tracking these metrics, you can better understand how different groups of users interact with the API and identify patterns of usage that can be used to improve the API and the user experience.

Monitoring usage metrics can help identify potential issues early on and take steps to fix them before they become bigger problems. This will help to improve the reliability of the API and ensure that it is providing a consistent and high-quality experience to the users. You will learn more about reliability metrics in the next section.

Reliability metrics

In the previous sections, you learned about the measurement of infrastructure performance and usage. Monitoring usage metrics helps to improve the reliability of an API by providing insights into how the API is being used and identifying potential issues or areas for improvement.

By tracking metrics such as the RPM, error rate, response time, and concurrent connections, you can identify patterns of usage and identify potential issues—such as high error rates, slow response times, or high concurrent connections—that may indicate that the API is not performing well or is being overloaded.

For example, if the error rate is high, it can indicate that there is a problem with the code, configuration, or infrastructure of the API. By identifying the cause of the errors, you can take steps to fix the problem and improve the reliability of the API.

By monitoring the response time, you can identify potential bottlenecks or issues with the API, such as slow database queries or a high number of requests. By identifying these issues, you can take steps to optimize the API, such as caching or indexing, or increasing the resources available to the API.

By monitoring concurrent connections, you can identify whether the API is being overloaded, and take steps to scale the API horizontally or vertically, to improve its reliability and performance.

In addition to performance and usage, you can also measure the reliability of your APIs to ensure that the performance and usage are consistent over time. Reliability metrics are key in ensuring that your APIs can be used by your customers for a prolonged period of time. The trust that your customers put into using your APIs will be the ultimate driver of customer satisfaction and enable more customers to trust your API offering.

There are five key API reliability metrics, as shown in the following screenshot:

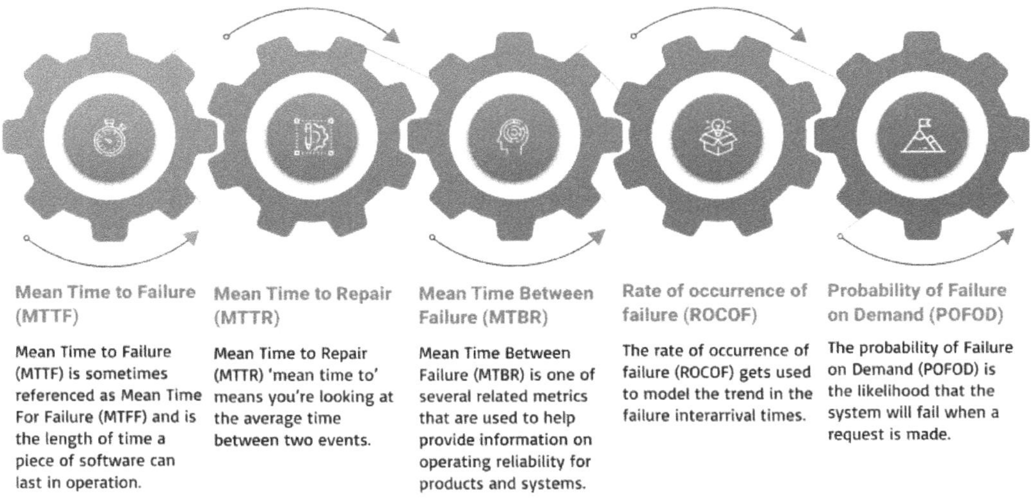

Mean Time to Failure (MTTF)
Mean Time to Failure (MTTF) is sometimes referenced as Mean Time For Failure (MTFF) and is the length of time a piece of software can last in operation.
Mean Time to Repair (MTTR)
Mean Time to Repair (MTTR) 'mean time to' means you're looking at the average time between two events.
Mean Time Between Failure (MTBR)
Mean Time Between Failure (MTBR) is one of several related metrics that are used to help provide information on operating reliability for products and systems.
Rate of occurrence of failure (ROCOF)
The rate of occurrence of failure (ROCOF) gets used to model the trend in the failure interarrival times.
Probability of Failure on Demand (POFOD)
The probability of Failure on Demand (POFOD) is the likelihood that the system will fail when a request is made.

Figure 10.7 – Key API infrastructure reliability metrics

In the following sections, you will learn in depth about each of these metrics and how to calculate them.

Mean Time to Failure (MTTF)

MTTF, also occasionally referenced as **Mean Time for Failure** (**MTFF**), is the length of time that a product lasts in operation. This metric is usually used for non-repairable resources and is widely used for hardware devices. MTTF helps measure how long your device or product can go on without running into a failure, and as the name suggests, a higher MTTF is a positive and desirable result. The following formula shows how MTTF is calculated:

$$\left[\frac{Totalhours of Operations}{Total number of Units}\right] = MeanTimeFailure(MTTF)$$

MTTF is measured across various testing conditions to determine how the product performs in different conditions. Metrics such as MTTF, **Mean Time to Repair** (**MTTR**), and **Mean Time Between Repair** (**MTBR**) are closely related and can be used to measure the reliability of your services. Based on analyzing these metrics, you can determine the right KPIs for your team and product. The following diagram shows the relationship between MTTF, MTTR, and MTBR:

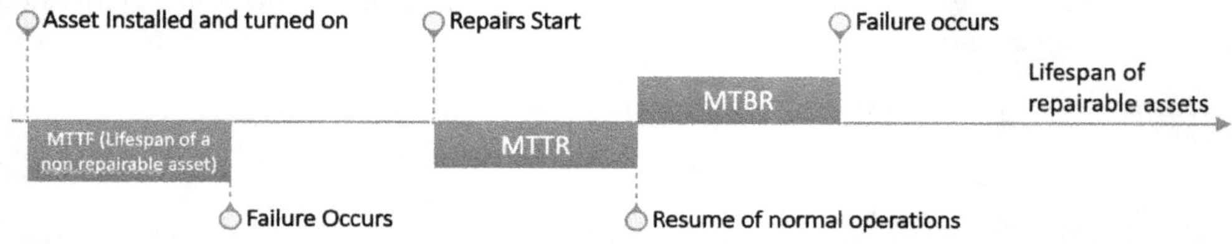

Figure 10.8 – Relationship between MTTF, MTTR, and MTBR

In order to understand MTTR and MTBR, it is important to understand MTTF as the foundational metric.

MTTR

As the name suggests, MTTR is the time between two events, the time from when repairs start, and the time when normal operations resume. The following formula shows how MTTR is calculated:

$$\left[\frac{Totalmaintenancetime}{Totalnumberofrepairs}\right] = MeanTimeRepair(MTTR)$$

MTTR is a crucial metric to measure how long it takes for your team in between incidents in terms of troubleshooting and fixing the necessary issues. If you have effective alerting tools and your team is able to fix issues quickly, you can keep the MTTR low and consequently prevent prolonged outages for your service.

MTBR

MTBR provides a way to measure the operating reliability for your products and systems by measuring the average time between failures. In some ways, this is a counter-metric to MTTR because even if your team is able to respond and repair services quickly, if you are experiencing frequent outages, that is still going to hurt reliability all the same. To ensure reliability, you must measure how quickly your team is able to repair a service but also be able to ensure the fixes are long-lasting so that outages can be avoided in the future.

The MTBR metric can be calculated using the following formula:

$$\left[\frac{Total\ operationaluptimebetweenfailures}{Numberoffailures}\right] = Meantimebetweenfailure(MTBR)$$

MTBR is also referred to as the average operating time between failures and helps measure how long your services are up on average between experiencing failures.

The Rate of Occurrence of Failure (ROCOF)

You learned about MTBR, which measures on average how long your services are up before experiencing failure. The ROCOF takes it a little further to identify trends in the failure interarrival times. For any repairable system, we want the system to fail infrequently and be repaired quickly, which translates to improving the ROCOF by increasing failure interarrival times.

A ROCOF of 0.02 means that there are likely to be 2 failures per 100 units of time. By conducting a statistical test on how often your services experience failure and plotting it against MTBR, you can determine statistically significant trends to understand the reliability of your infrastructure.

However, most of the time , you may not have enough data because your services may be new, and you may be yet to collect the data required to do such an analysis. You should still plan to start collecting this data in order to eventually be able to measure the ROCOF as this metric can be very insightful in measuring the reliability of your system and help you set better KPIs.

Probability of Failure on Demand (POFOD)

POFOD, also known as **PFD**, is the probability of a system failing when a request is made. A PFD rate of 0.001 means that 1 in every 1,000 requests may result in a failure.

When it comes to safety-critical systems, POFOD is an absolutely necessary safeguard. POFOD is applicable to security systems in which the provision of services is required only occasionally.

Summary

As a product manager, you may not be responsible for instrumenting infrastructure metrics. But you can partner with the engineering teams to instrument and track the necessary metrics and make sure these are prioritized on the engineering roadmap. Having insight into the reliability, performance, and usage metrics atthe level of infrastructure ensures that your product is strong at a foundational level. Ultimately, the product experience and business are dependent on the reliability of your product, and any efforts to grow your product are futile if the infrastructure fails your customer experience.

Once you have a functional and reliable API, the next step is to ensure a good customer journey that allows customers to discover and start using your APIs. In the next chapters, you will learn about product metrics that measure various aspects of the customer journey and identify opportunities to streamline the experience.

11
API Product Metrics

In the previous chapter, *Infrastructure Metrics*, you learned about the infrastructure metrics that form the foundation of your API analytics. Assuming your APIs are operational and in a position to meet customer expectations, you can start thinking about product metrics and start measuring the your customers' journey across discovery, activation, acquisition, retention, and the overall experience. By keeping track of every step a customer takes, you can find out where they get stuck, where they fail, and what your developer experience does best.

In this chapter, you'll find out about the different metrics you can use to learn more about your customers. The main topics covered in this chapter include the following:

- Defining product metrics
- Discovery
- Engagement
- Acquisition
- Activation
- Retention
- Experience

The metrics you will learn about in this chapter can be used across all the stakeholders in your product to align on common goals and priorities.

Defining product metrics

Product metrics are used to figure out how well a product is doing by measuring different parts of the customer journey and comparing them to the product strategy. Product metrics enable you to identify opportunities for product development, establish a baseline, and evaluate results as you make changes to the overall product experience. To accomplish this, product metrics measure various aspects of the customer journey, including discovery, engagement, activation, acquisition, retention, and experience.

You can use the developer journey you learned about in *Chapter 5, Growth for API Products,* to map the steps of the developer journey to various metrics you can measure at each step as customers go through them. You can see this framework in the following figure:

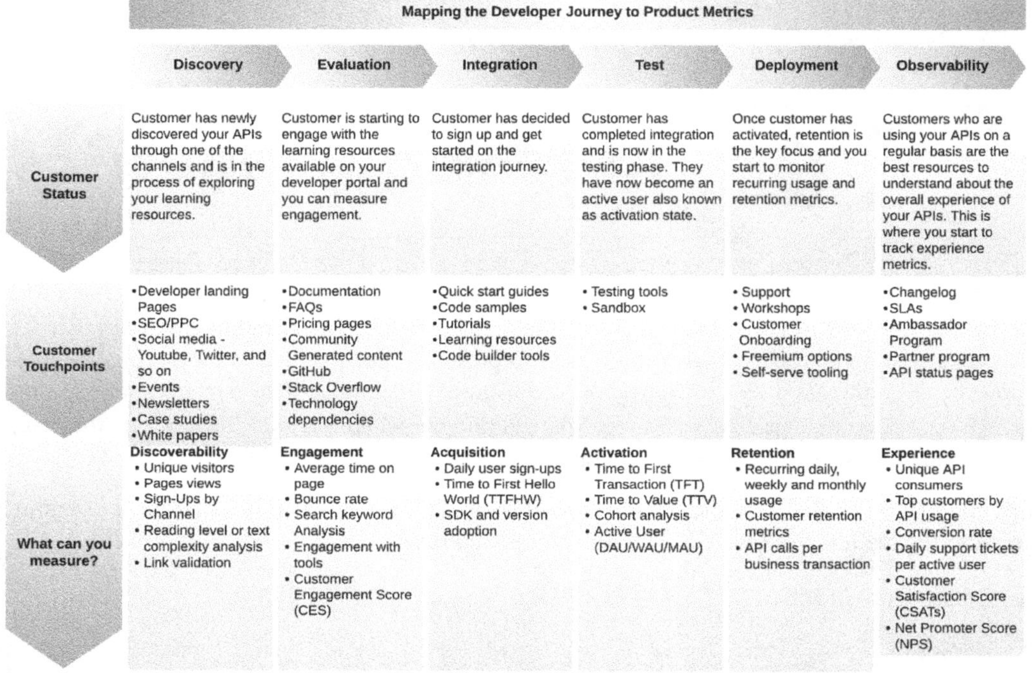

Mapping the Developer Journey to Product Metrics
Discovery
Evaluation
Integration
Test
Deployment
Observability
Customer Status
Customer has newly discovered your APIs through one of the channels and is in the process of exploring your learning resources.
Customer is starting to engage with the learning resources available on your developer portal and you can measure engagement.
Customer has decided to sign up and get started on the integration journey.
Customer has completed integration and is now in the testing phase. They have now become an active user also known as activation state.
Once customer has activated, retention is the key focus and you start to monitor recurring usage and retention metrics.
Customers who are using your APIs on a regular basis are the best resources to understand about the overall experience of your APIs. This is where you start to track experience metrics.
Customer Touchpoints
•Developer landing Pages •SEO/PPC •Social media - Youtube, Twitter, and so on •Events •Newsletters •Case studies •White papers
•Documentation •FAQs •Pricing pages •Community Generated content •GitHub •Stack Overflow •Technology dependencies
•Quick start guides •Code samples •Tutorials •Learning resources •Code builder tools
• Testing tools • Sandbox
• Support • Workshops • Customer Onboarding • Freemium options • Self-serve tooling
•Changelog •SLAs •Ambassador Program •Partner program •API status pages
What can you measure?
Discoverability • Unique visitors • Pages views • Sign-Ups by Channel • Reading level or text complexity analysis • Link validation
Engagement • Average time on page • Bounce rate • Search keyword Analysis • Engagement with tools • Customer Engagement Score (CES)
Acquisition • Daily user sign-ups • Time to First Hello World (TTFHW) • SDK and version adoption
Activation • Time to First Transaction (TFT) • Time to Value (TTV) • Cohort analysis • Active User (DAU/WAU/MAU)
Retention • Recurring daily, weekly and monthly usage • Customer retention metrics • API calls per business transaction
Experience • Unique API consumers • Top customers by API usage • Conversion rate • Daily support tickets per active user • Customer Satisfaction Score (CSATs) • Net Promoter Score (NPS)

Figure 11.1 – Product metrics interpreted from the developer journey

As customers discover your APIs, there are a number of discoverability-related metrics you can start to measure. As customers move from discovery to evaluation, they begin to engage with the tools you provide. You can measure the engagement rates for these tools to understand how customers are able to use them. Once customers sign up and become active users, you can start to use their usage patterns and start tracking activation, retention, and overall experience metrics over time.

Product metrics lead to effective decision-making by enabling you to confirm your hypotheses, communicate your insights with the team and leadership, and ultimately, help align stakeholders. Gathering supporting data for any hypothesis allows you to get approval and buy-in from leadership for the initiatives you propose.

When you have a variety of metrics in your product metrics strategy, you can choose the one that best measures the product goal you are working on at any given time. This metric is going to be your north star. The north star metric is the one you want to influence in the short term and should be linked to your business goals. The following diagram shows the six types of product metrics and the different types of metrics within each type.

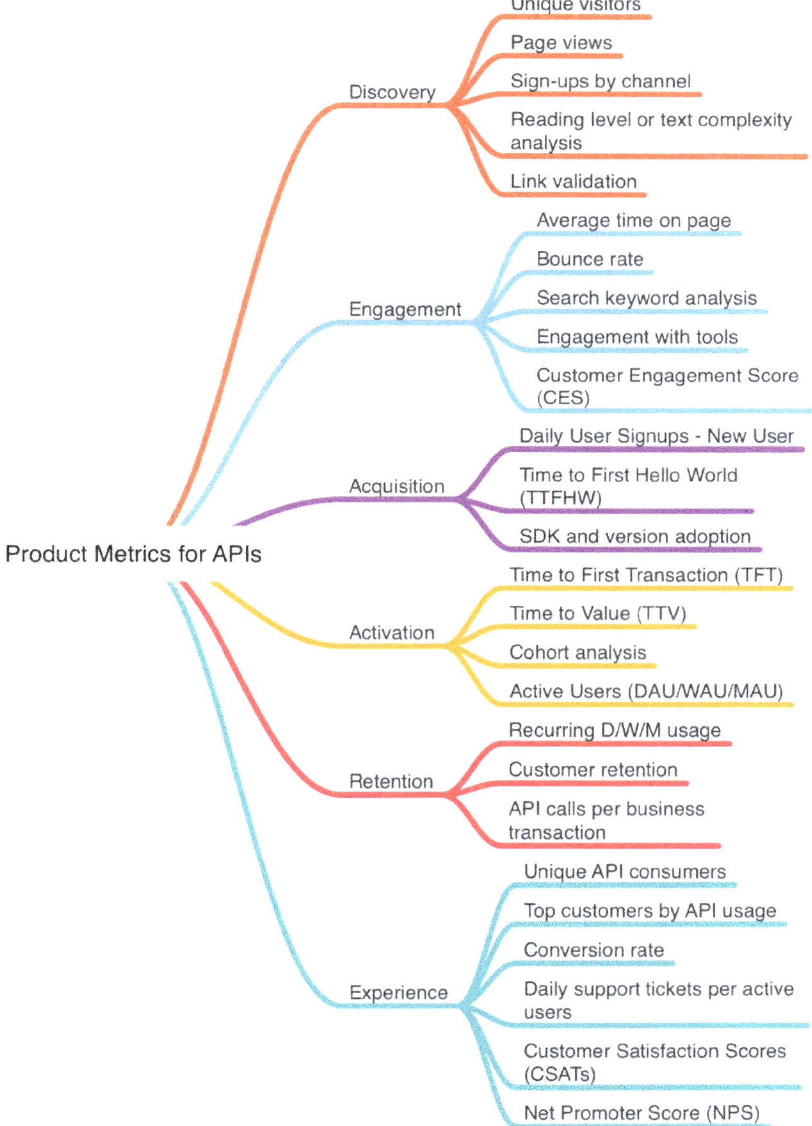

Figure 11.2 – Product metrics across discovery, engagement, acquisition, activation, retention, and experience

Let's take a look at these metrics in detail:

- **Discovery**: Customers can discover your APIs from various channels, such as social media and organic searches. Discovery metrics help measure the efficiency of the channels that drive discovery for your consumers.
- **Engagement**: Once customers discover your developer portal and explore the capabilities and tools you offer to integrate with your APIs, you will start to see engagement signals in terms of how much time they spend on these pages and their interactions with the tools.
- **Acquisition**: Customers must experience the product at least once to be able to evaluate the value it can drive for them appropriately. The first successful API calls are a great signal to measure customers arriving at this point in the journey.
- **Activation**: The definition of activation may vary from organization to organization and needs to be established based on your business goals. According to your definition of an active user, metrics that measure the customer journey to being considered active are categorized under activation metrics.
- **Retention**: All the effort you put into gaining customers is only useful if customers continue to use your product for a prolonged time. Customer retention metrics allow you to gain insights into when and how you lose customers.
- **Experience**: Throughout the customer journey from discovery to retention, customers go through various stages of decision-making, integration, testing, and so on. At each stage, they rely on the experience you design and the support channels you provide to help them through the journey. Measuring the overall experience with metrics such as the **Customer Satisfaction Score** (**CSAT**) and **Net Promoter Score** (**NPS**) allows you to keep a close eye on how customers perceive the overall product experience.

We begin establishing these metrics using the time it takes for customers in each phase of interaction with the product. The average time spent by the user at each phase sets your current state or baseline metrics. If your APIs are intuitive and the documentation and tooling are easy to navigate, customers will be able to learn about them and make decisions fairly quickly.

You can use the same methodology to walk through competitor products and do a competitive benchmarking of your customer journey. This will enable you to see how you compare to your competitors in terms of the number of steps in the customer journey and the overall time-to-value TTV for your customers.

You can observe a steep drop-off in usage patterns, users taking longer than expected in certain parts of the customer journey, and variations in time spent on product features. In the following sections, you will learn about each of the categories of product metrics and the metrics under each of them.

Discovery

Web analytics on your developer web pages can tell you which pages send the most traffic to your site and how easy it is for customers to find their way around and use the information you give them. The following diagram shows the metrics that you can use to measure the discoverability of your APIs:

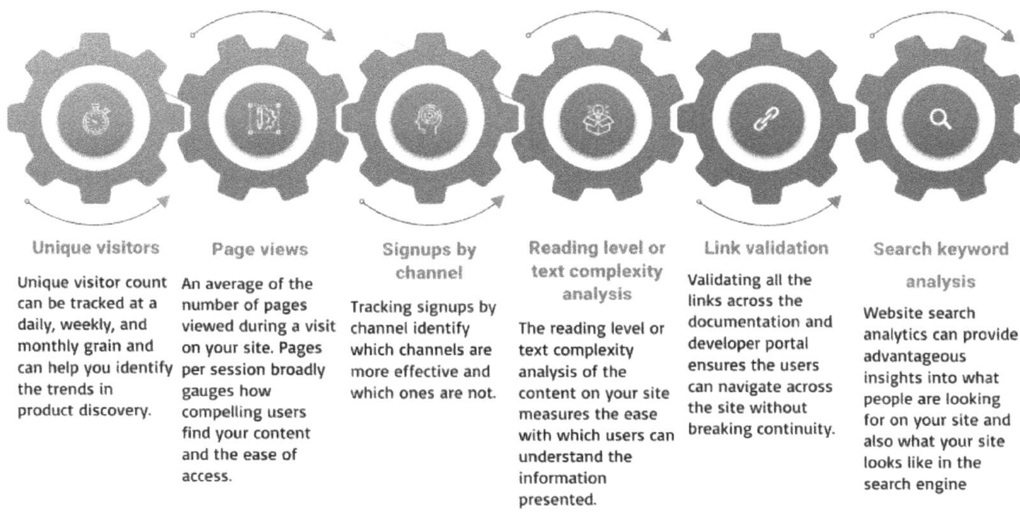

Unique visitors
Unique visitor count can be tracked at a daily, weekly, and monthly grain and can help you identify the trends in product discovery.
Page views
An average of the number of pages viewed during a visit on your site. Pages per session broadly gauges how compelling users find your content and the ease of access.
Signups by channel
Tracking signups by channel identify which channels are more effective and which ones are not.
Reading level or text complexity analysis
The reading level or text complexity analysis of the content on your site measures the ease with which users can understand the information presented.
Link validation
Validating all the links across the documentation and developer portal ensures the users can navigate across the site without breaking continuity.
Search keyword analysis
Website search analytics can provide advantageous insights into what people are looking for on your site and also what your site looks like in the search engine

Figure 11.3 – Product metrics to measure the discoverability of your APIs

In the sections that follow, you will learn more about the different metrics that measure different parts of how easy it is to find your APIs.

Unique visitors

Unique visitors on your developer site are counted as any users who have visited your site for the first time and will not be recounted if they visit again. Most web analytics tools, such as Google Analytics and Heap Analytics, provide the ability to get unique visitor counts for any web page as a starting point for web analytics.

Unique visitor counts can help you figure out whether your site is interesting and easy to find. They can also be used to compare your site to industry standards and competitors. This metric also lets you track how well a campaign is doing and measure how well inbound marketing is doing.

Page views

Another metric that most web analytics tools provide out of the box is page views. This metric measures the number of pages viewed on a site. A high number of page views could mean that people are spending a lot of time on the site and are interested in the content, but it could also mean that they can't find what they're looking for. To understand the page views metric, you need to compare it to other metrics, such as the number of sessions.

Sign-ups by channel

By keeping track of developer sign-ups by referral channel, you can see which channels are best at bringing developers to your site. To get people interested in your APIs, the sales, marketing, and developer relations teams must always make content, host events, organize workshops, and so on. This interest must translate into developers visiting and signing up for your developer portal. Using web analytics tools, you can track the referral links to sign-up data and determine which channels are effective and which aren't.

By figuring out how well different channels work, your team can set priorities and come up with a plan to get your APIs in front of the right people.

Reading level or text complexity analysis

API documentation can be very text-heavy, and this results in the need to be thoughtful about how complex the text is and whether it meets the reading level of the audience. As you learned in *Chapter 7, Walking in the Customer's Shoes*, there are several customer personas that will be part of the customer base for your APIs. Customer personas such as product managers, developers, software architects, and security engineers might read your documentation. Making sure your content is accessible to these different personas, easy to navigate, and meets their different levels of technical knowledge is critical to making your documentation and tooling usable.

There are several quantitative dimensions of text complexity. You can measure the levels of meaning in the text, which could be a single meaning or multiple meanings. Using clear language that says what a piece of text is about in a way that leaves no room for doubt is always helpful to users.

The most important part of making a good user experience is how your documentation is organized. Your UX design partners can help you figure out the best way to organize the information in your documentation and developer portal so that it's easy for customers to use. Several tools, such as ReadMe, help you generate API references directly from API specifications and create standardized API references with all the functionality you need to get your API documentation off the ground. Such tools are great for early-stage products, but you might want to build solutions in-house as your use cases become more complex.

Use simple, consistent language throughout the entire customer experience to keep a user's mind from having to work too hard. This requires a conscious effort to curate the terminology that is used. Terminology can be categorized into the following:

- **Engineering terminology**: These are terms that are common to the technology space and not specific to your product. To a certain degree, you can expect developers to know the fundamentals of the technology, but it is still a good practice to link the knowledge base wherever possible. An example of engineering terminology is when you link the term *REST APIs* to a Wikipedia article that describes what RESTful APIs are.
- **Domain-specific terminology**: These are terms that are specific to your product's business domain. In the case of financial products, this could be financial terms such as *debit* and *credit*. You cannot expect all developers to be familiar with these terms. You should ensure that your documentation helps developers of all knowledge levels to grasp these terms in the context of your APIs.
- **Product-specific terminology**: As you design your APIs and the tooling to support them, you will have to create terminology that goes along with them. You should be mindful of the terminology you use to describe your product to ensure that it doesn't contradict any existing engineering or business terms.

It might be impossible to keep track of all the terminology used across all the tools and documentation. You can still create guidelines for nomenclature across the teams to ensure the best practices are followed. Keeping the language simple so that developers of all skill levels can use your APIs and ensuring the terminology is clear and unambiguous will make your APIs easier to use, as well as make it easier for customers to get started with them.

Link validation

Over time, your API documentation is bound to grow. There will soon be hundreds, if not thousands, of web pages on your site linked to each other, creating a complex web of links. From an SEO perspective, this is both expected and desired. When you link to things outside of your site, such as videos, blogs, and social media posts, these *backlinks* will make it easier for people to find your APIs.

The challenge that accompanies this complex web of links is that, over time, links can break. If you have a blog that links to a specific page in your documentation but that link needs to be fixed, the customer is effectively unable to move from the blog to your site and move forward in their journey. Broken links on your site are ultimately broken customer journey points. You should be tracking links across your site to ensure that your customers don't run into invalid or broken links as they go through your content.

There are several tools that offer link validation functionality, such as **Sitechecker**, **SE Ranking**, **Semrush**, and **W3C Link Checker**. These are usually part of an SEO analysis. The documentation for an API works like any other content when it comes to searchability, and all SEO best practices apply. This includes keyword optimization, URL validation, and analysis of backlinks. Most SEO tools offer these features right out of the box. You can check these stats regularly to ensure that your site's most important parts are working and that your customers can find and use your documentation easily.

Search keyword Analysis

When people use search engines to find your website, you can learn a lot about how to improve your rankings and how to spend your advertising budget by conducting a keyword study. Therefore, search marketing efforts should begin with a thorough keyword study. Search keyword analysis can provide you with important insights into the keywords that drive traffic to your product but also how your competitors are positioning their offerings.

When you do keyword analysis, think about the use cases your APIs solve for your customers and identify the difference between business terms, domain-specific terms, and technical terms. The decision-makers on the consumer side are often more likely to look for business terms while they search for solutions to their business problems. Making sure that you optimize your site for such terms will result in the right audience being able to find your APIs.

Engagement

Once your customers discover your APIs and start getting interested in using them, they will likely spend more time on your developer documentation and start interacting with the tools you provide them with to help evaluate your APIs.

Customer engagement is the level at which a customer is interested in your product. Several indicators related to customer engagement can be used to assess this. By calculating customer engagement, you can get a wealth of useful insights that you can use right away. Your website, social media pages, forums, chatbots, email newsletters, blogs, videos, third-party websites, and so on can all serve as entry points for this engagement.

The following diagram shows a few metrics you can use to measure engagement across your developer experience:

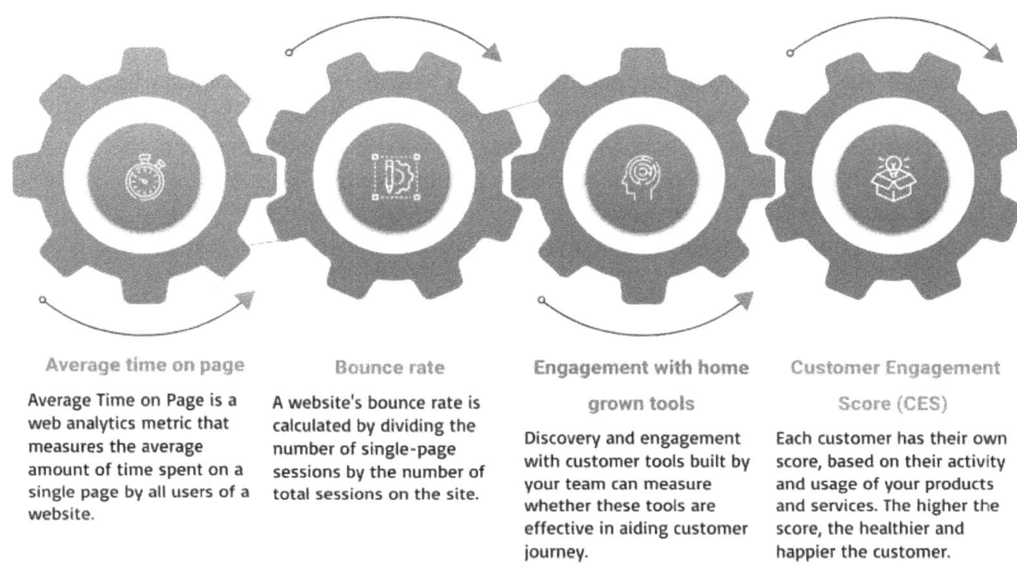

Figure 11.4 – Product metrics to measure customer engagement

In the following sections, you will learn about the average time on page, bounce rate, search keyword analysis, engagement with homegrown tools, and the **Customer Engagement Score** (**CES**) that form the customer engagement metrics.

Average time on page

The average amount of time people spend on a given page is calculated using the web analytics metric known as average time on page. This metric tracks the average amount of time users spend on pages that don't lead them to abandon the site. A website's *exit page* is the page from which a user leaves the site. The formula for calculating the average time on a page is shown here:

$$\left[\frac{Sum(Time\ Spent\ on\ Page)}{(\ Count(Page\ Views) - Count(Page\ Exits)\ )}\right] = Average\ Time\ on\ Page$$

A page with 1,000 page views, 800 page exits, and 100 minutes of time spent on it has an average time spent of 100/(1,000 – 800) = 100/200 = 0.5 minutes or 30 seconds.

An industry-wide benchmark for average time on a page across industries is 52 seconds, but this number can vary based on device types, industry, and the nature of the audience.

Bounce rate

The percentage of users that see only one page on your site before leaving (or *bouncing* back) to the search results or the referring website is known as the bounce rate. If a user is idle on a page, that can also be counted as a bounced session. A very low bounce rate isn't necessarily good, and a bounce rate of 25% or lower usually means something is broken. A bounce rate of 24%–40% is considered ideal, while a bounce rate of up to 55% is considered average. A bounce rate of 56% or higher is considered high and something that would need to be analyzed further to improve the site experience.

Site-loading speed has a big impact on bounce rates. If your site takes longer than expected to load, users might lose interest by the time it loads and leave the site. **Google PageSpeed Insights** and the **Google Search Console PageSpeed reports** are common tools that provide insights into site-loading speeds.

If your content is self-service, users might be able to find the content they need quickly, and that may be the reason for a high bounce rate. Dictionary sites are a great example of sites with high bounce rates, which are also complemented by shorter average session durations.

Engagement with homegrown tools

It is common for companies to build custom tools that allow customers to generate code or get a demo of the API integrations. These tools help users imagine and evaluate how they might use your APIs, which cuts down on the time it takes to make a decision.

As customers engage with the tools you provide, you can gather data about their interactions and learn about what they are looking for. The best thing about custom tools is that you can get very detailed information about how customers interact with them. This can help you understand how customers think and help you to meet their needs better.

A great example of custom tooling provided by a company is the credit card generator provided by PayPal, which enables customers to generate test credit cards for the purpose of processing transactions in a sandbox environment. PayPal's sandbox credit card generator is part of the developer portal shown here:

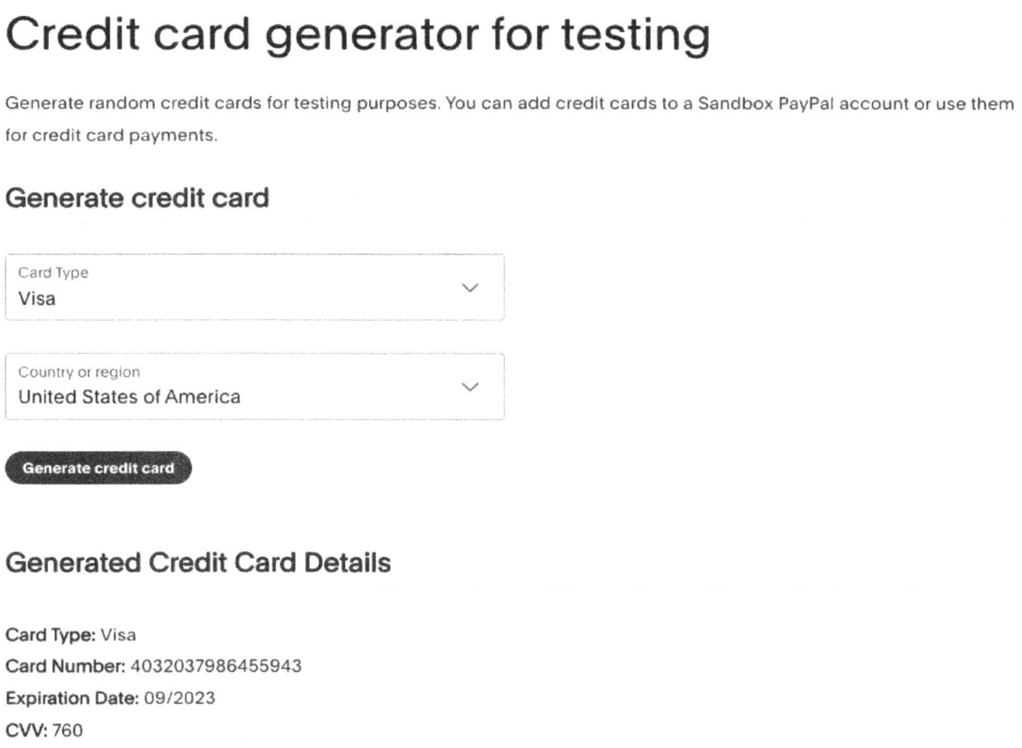

Credit card generator for testing

Generate random credit cards for testing purposes. You can add credit cards to a Sandbox PayPal account or use them for credit card payments.

Generate credit card

Card Type
Visa

Country or region
United States of America

Generate credit card

Generated Credit Card Details

Card Type: Visa
Card Number: 4032037986455943
Expiration Date: 09/2023
CVV: 760

Figure 11.5 – PayPal's credit card generator for testing, available on PayPal's developer dashboard

From the usage data analysis of tools such as the credit card generator shown previously, you can get data on what kind of use cases customers are evaluating as part of their integration. If customers run into issues or create support requests for additional types of credit cards to be supported, you will also get insights into what your product offering might be missing that you could prioritize for development.

Customer engagement score (CES)

As the name suggests, the CES is a measure of how engaged your customers are. Engagement is a strong signal of how customers feel about your product, and along with the CSAT, the CES can provide valuable insights into which customers are ready to purchase, which customers are about to churn, and finally, when it is a good opportunity to upsell to an existing customer.

The definition of engagement is not limited to just page views, session durations, or pages per session. When we think about customer engagement more holistically, we have to consider the value customers get from a product or service as the key variable. In the case of APIs, a customer can potentially onboard really quickly if they use some copy-and-paste code. They might be able to start making their first API calls and even scale their usage in very little time. However, this can only be interpreted as value when they find the service truly valuable.

A customer who is just starting out on their customer journey might find it useful to be able to use your Postman Collections to build their application. But if that doesn't lead to the customer building a production application that gets used often, then the customer engagement at the collection level isn't worth much.

The first step toward measuring the CES is to make a list of the key benefits of your APIs. For payment APIs, the key benefit is being able to accept payments. For email APIs, the main benefit to the customer is being able to build applications that send and receive emails. You need to track the value you provide to your customers in terms of specific events, actions, and results that can be tracked.

Once you have identified the events or actions that drive value for your customers, you can prioritize them and assign weights to them. The greater the benefit of an interaction, the higher value you can assign to that interaction. Using a scale of 1 to 100 is beneficial because there is less chance that you will ever use 100 or more variables to compute the CES.

The most common way to compute customer engagement scores is the following:

$$\textit{Customer Engagement Score} = (w1 * n1) + (w2 * n2) + \ldots + (w\# + n\#)$$

Here, *w* is the weight given to an event or interaction, and *n* is the number of times the event occurs. When you compute the CES over time, you can begin to see clusters of customers that might be at risk of churn, need product training, or struggle with tooling. These patterns can help guide your product and sales strategy and help you build better roadmaps that deliver more value to the customers.

Acquisition

As a user starts to engage with the APIs and starts to make their first API calls, you would start to see sign-up data and usage metrics such as **time to first hello world** (TTFHW). New user acquisition is a key metric for all teams that are working on growth-related initiatives, such as marketing and developer relations. In the following diagram, you can see the primary acquisition metrics you will learn about in this section:

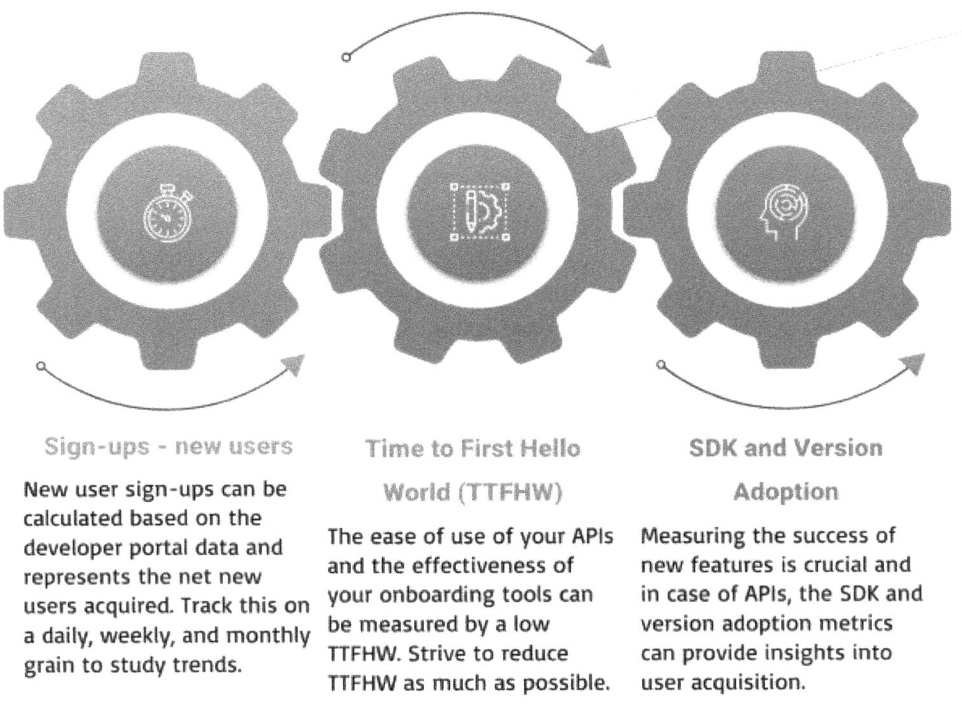

Sign-ups - new users
New user sign-ups can be calculated based on the developer portal data and represents the net new users acquired. Track this on a daily, weekly, and monthly grain to study trends.
Time to First Hello World (TTFHW)
The ease of use of your APIs and the effectiveness of your onboarding tools can be measured by a low TTFHW. Strive to reduce TTFHW as much as possible.
SDK and Version Adoption
Measuring the success of new features is crucial and in case of APIs, the SDK and version adoption metrics can provide insights into user acquisition.

Figure 11.6 – Product metrics for measuring acquisition

In the following sections, we will dive deeper into each of the acquisition metrics.

Daily user sign-ups – new users

The user journey for your APIs begins with users discovering them through social media, blog posts, video tutorials, and so on. However, this can often be a much broader set of audiences that are not at a point of decision-making, where they are ready to start using your APIs. Users who complete the discovery process and start engaging with your product must then sign up to use it. Signing up on the developer portal is the first sign of true interest from a user. In the following diagram, you can see the user journey from sign-up to making their first API call, which are acquisition metrics, followed by time to first transaction and time to value, which are activation metrics that we will see in the next section.

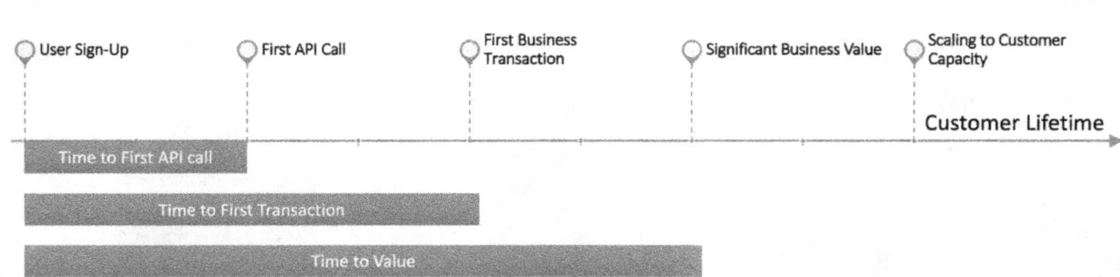

Figure 11.7 – The user acquisition and activation journey starts at sign-up and goes through the first API call, business transaction, and customer lifetime value

Tracking daily user sign-ups provides insights into new user onboarding on your developer portal. You can measure this either by raw account creation data. Sign-ups do represent an important moment in the user journey, but you don't have to stop there. You can also measure the number of users who were able to find and copy API credentials from their developer accounts. By measuring credential discovery in addition to user sign-ups, you can figure out how good the sign-up action is. If users create developer accounts but don't even get their credentials, it could be because they can't find them or don't want to. Both of these can make it harder for the user to keep using your APIs and should be looked into more to improve the user experience.

Time to first hello world (TTFHW)

The TTFHW measures how long it takes for a user to go from discovering your APIs to making their first API call. The sign-up time is often used as the starting time for this calculation because it is usually the easiest to get, but if you can go further back to track the user from discovery, that would be more helpful in measuring the entire journey from the first API call for a user.

The TTFHW is a key metric that all API product managers should prioritize because it measures the entire developer journey, up to the first "*aha*" moment for the customer when they experience your API for the first time. In *Figure 11.6*, the TTFHW metric maps the customer journey as part of the customer's lifetime value.

You can either measure the TTFHW as a single average metric across all customers or you can segment it based on acquisition channels. Certain channels might have a quicker journey for a customer to reach their first API call, such as a Postman Collection. If you publish a public Postman Collection, you will enable customers to discover your APIs and make their first APIs calls in a matter of minutes because they would not have to write any code. However, this might not be the case for all channels.

If you segment the TTFHW metric across different acquisition channels for your customers, you will discover which channels are more effective at getting customers to make their first API calls with less difficulty, helping you also to identify opportunities to improve other channels.

Software development kit and version adoption

Software development kits (**SDKs**) facilitate your customers' usage of your APIs. API teams often manage tens or hundreds of SDKs that support different features, and this can be challenging from a maintenance point of view. Popular SDKs are kept up to date, while those less frequently used are not – a telling sign for API usage.

Measuring the adoption of various SDK versions can give insights into which versions need to be maintained and which can be deprecated. You can also dive into the usage to understand which endpoints and features are being used by users, based on the various SDK versions that are popular.

Activation

Activation metrics attempt to measure how many users have started to use your product and find value. The first challenge of defining an activation metric is how you define a user as being active. This usually varies from business to business. For example, if you are looking at a business in the payments domain, you can expect that an active user will have a certain number of weekly transactions. Similarly, if you look at the communications domain, a user sending a minimum of a certain number of messages a day might be considered active. Any users who are actively using the product can be considered active users.

The following diagram shows a few of the key activation metrics you will learn about in this section:

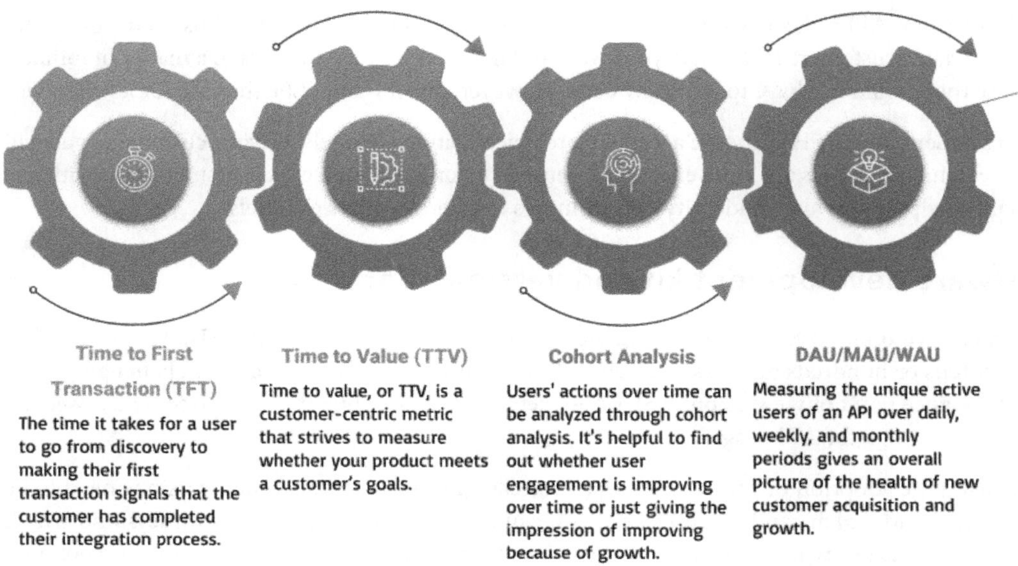
Time to First Transaction (TFT)
The time it takes for a user to go from discovery to making their first transaction signals that the customer has completed their integration process.
Time to Value (TTV)
Time to value, or TTV, is a customer-centric metric that strives to measure whether your product meets a customer's goals.
Cohort Analysis
Users' actions over time can be analyzed through cohort analysis. It's helpful to find out whether user engagement is improving over time or just giving the impression of improving because of growth.
DAU/MAU/WAU
Measuring the unique active users of an API over daily, weekly, and monthly periods gives an overall picture of the health of new customer acquisition and growth.

Figure 11.8 – Product metrics to measure activation

The following subsections will dive deeper into the individual metrics – **time to first transaction** (TFT), **time to value** (TTV), cohort analysis, and daily, weekly, and monthly active users (DAU, WAU, and MAU respectively) – that are used to measure activation.

Time to first transaction (TFT)

The time it takes for a user to go from discovery to making their first transaction is called time-to-first-transaction, or TFT. This is a key metric because it signals that the customer has completed their integration process. The definition of the transaction here is a production API call.

In most organizations, there is a test or *sandbox* environment and a production environment. Users might make a number of API calls in the sandbox environment as part of the development process. However, when their integration is ready and their application is in production, they will start to make production API calls. This is best understood with examples such as PayPal, where, in a sandbox environment, users can make API calls that would not move any money in real bank accounts; the transactions would only show in sandbox environments that simulate the production experience. However, once a customer application is in production, there are real banking transactions being processed.

Calculating how long it takes for a user to first start with the APIs and get to their first transaction allows you to measure the entire journey of the user, from learning about the APIs to successfully integrating with them. Depending on the complexity of the application, the size of the development team, and the ease of use of your APIs, the users might take anywhere from a few weeks to several months to complete an integration successfully.

The goal of the TFT metric is to understand how long it takes for your customers to integrate with your APIs. Sort the TFT data into groups to find out whether there are users or use cases in which some customers can integrate much faster than others. This will help you identify the ideal customers and potential product-market fit if you haven't already established it.

Time to value (TTV)

Time to value, or TTV, is a customer-centric metric that strives to measure whether your product meets a customer's goals. Value in this case is defined as the benefit to the customer. When a customer successfully integrates with a set of APIs, they are looking for an easy onboarding experience that allows them to start making production API calls quickly and go live with their application. This lets them get business value as soon as their app starts to use your APIs properly. Depending on the domain of your APIs, the number of API calls that can be considered valuable to the customer might vary. You should look at how your users use your product and talk to them to find out what they think is the first sign of value.

You can define value as the first business transaction in domains such as finance. If a customer has started using your APIs to process payments, even the first payment is a good sign of value. You can also define value in terms of reaching a level of usage – for example, reaching 100 API calls a week.

Once you establish the definition of value, you can then start to analyze your user base against this metric better to understand the cohorts and behavior patterns of your users. The TTV metric lets you measure and figure out how long it takes for a user to start seeing the value of your APIs. Divide the TTV metric into different customer personas to find out whether some users see the value of your product or service sooner than others.

The TTV might shift based on the type of customer, the type of business, and the services being provided. When a customer is valuable to you, it's not the same as when you are valuable to them. Throughout the sales cycle, it is crucial to remember the customer's top objectives. The TTV of your customers is something you should aim to reduce gradually. A low TTV for a business is a good indicator of its future success.

Cohort analysis

Users' actions over time can be analyzed through cohort analysis. It's helpful to find out whether user engagement is improving over time or just giving the impression of improving because of growth.

Cohort analysis is a methodology in behavioral analytics that helps you study users who have similar experiences. Instead of looking at all the users of a platform, web app, or online game as a single entity, the data is analyzed by making subsets of users with similar traits. Cohorts are groups of people with something in common or similar synchronous experiences.

Cohort analysis is helpful because it makes it easy to tell the difference between growth and engagement data. This is because growth can sometimes hide engagement problems. In reality, the huge number of new users is hiding the fact that only a small portion of the original users are still using a site.

The following diagram shows an example of a cohort table for an app over a 10-day period, comparing the cohorts of users as they onboard:

App launched	% Active users after app launch											
Cohort	Users	Day 0	Day 1	Day 2	Day 3	Day 4	Day 5	Day 6	Day 7	Day 8	Day 9	Day 10
January 25	1100	1100	27.50%	26.40%	19.80%	18.70%	17.60%	15.40%	15.40%	13.20%	14.30%	13.20%
January 26	984	984	25.58%	19.68%	16.73%	16.73%	15.74%	13.78%	14.76%	11.81%	12.79%	
January 27	678	678	23.05%	16.27%	12.20%	10.85%	10.85%	10.17%	9.49%	9.49%		
January 28	535	535	16.59%	12.84%	9.63%	9.10%	8.03%	8.03%	7.49%			
January 29	1010	1010	28.28%	18.18%	16.16%	16.16%	16.16%	15.15%				
January 30	908	908	24.52%	21.79%	16.34%	14.53%	13.62%					
January 31	1191	1191	41.69%	23.82%	19.06%	19.06%						
February 1	621	621	19.25%	15.53%	0.09936							
February 2	630	630	21.42%	14.49%								
February 3	922	922	32.27%									
All users	8579	100	26.01%	18.78%	14.98%	15.02%	13.67%	12.50%	11.79%	11.50%	13.55%	13.20%

Figure 11.9 – Cohort analysis showing active user behavior over user acquisition cohorts

The preceding cohort table provides two key benefits; by comparing different cohorts at the same point in the product's life cycle (as shown vertically downward in the table), we can determine what percentage of users in a certain cohort are returning to the app after 3 days, and so on. User lifetime, which is shown horizontally to the right of the table, tells you how long people keep coming back and how strong or valuable that cohort is. The success of the first few months of a user's time with your product depends on how well your onboarding process and how effective your customer success teams are at helping new users. This probably has to do with things such as product quality, efficiency, and responsiveness to consumer concerns.

DAU/MAU/WAU

Measuring the unique active users of an API over daily, weekly, and monthly periods gives an overall picture of the health of new customer acquisition and growth. The terms DAU (daily active users), WAU (weekly active users), and MAU (monthly active users) are often used to describe the number of active users on a daily, weekly, and monthly basis, respectively. Many teams compare the API MAU to the web MAU to show full product health.

Once you are able to establish the DAU, WAU, and MAU metrics, you can start to analyze them based on cohorts such as acquisition channels. When a customer signs up to use your API, you should also be tracking the referring domain and UTM parameters. This lets you group weekly active users by

UTM source or UTM campaign so that you can learn more about the marketing channels that lead to increased usage and engagement.

The following diagram shows you the **Weekly active tokens by acquisition channe**l dashboard created for APIs by the **Moesif** API analytics tool.

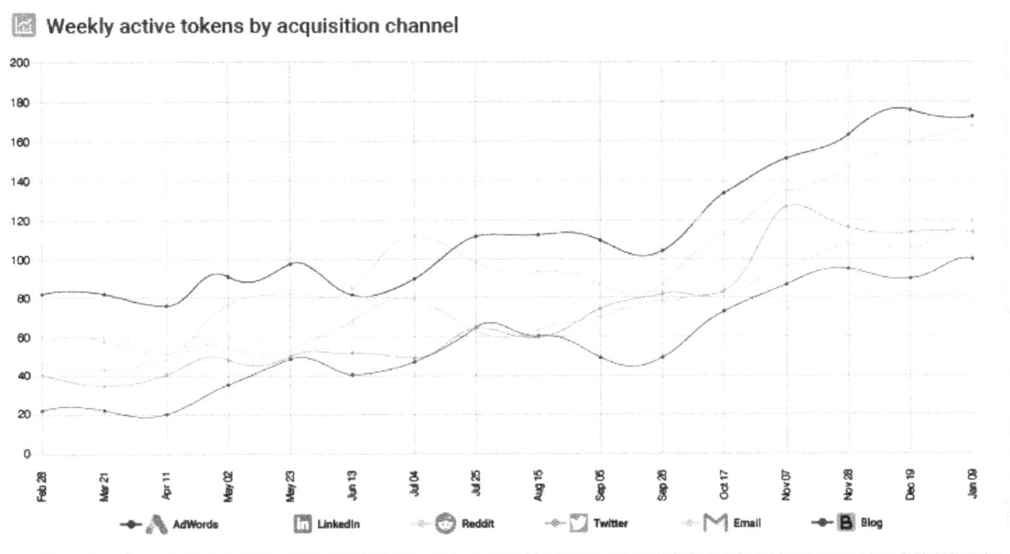

Figure 11.10 – Moesif's dashboard for weekly active tokens by acquisition channel

The preceding diagram shows how you can track weekly active users by acquisition channel using an API analytics tool such as Moesif. This would give you an understanding of how users acquired by different channels vary in their activity and engagement with your APIs.

Now that you have learned about individual metrics for TFTTTV, cohort analysis, and DAU, WAU, and MAU metrics that are used to measure activation, in the next section, you will learn about user retention.

Retention

While you spend so much time and effort in helping customers discover and engage with your products, the effort would be a waste if, for any reason, they were not loyal to the product and you were not able to retain them as customers for a prolonged period of time. Loyalty over time is a great measure of the overall success of your product, and that is why customer retention is a key product metric that all types of products must track.

The following diagram shows the retention metrics you will learn about in the following sections:

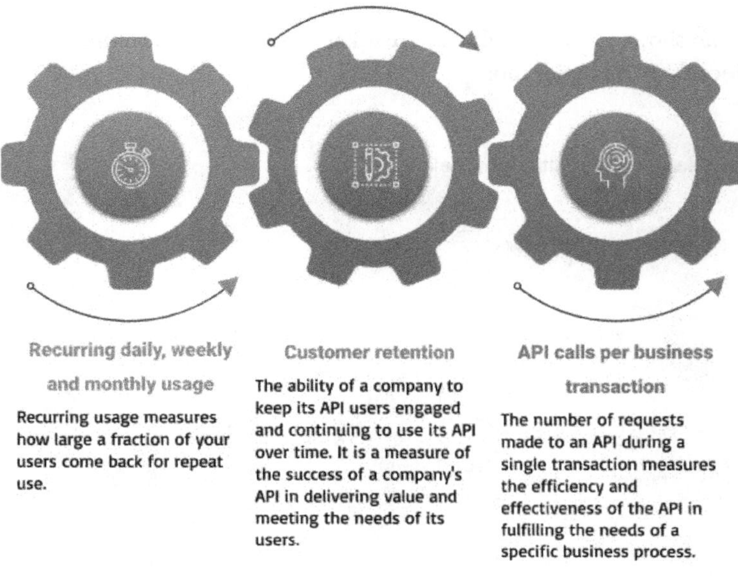

Recurring daily, weekly and monthly usage
Recurring usage measures how large a fraction of your users come back for repeat use.
Customer retention
The ability of a company to keep its API users engaged and continuing to use its API over time. It is a measure of the success of a company's API in delivering value and meeting the needs of its users.
API calls per business transaction
The number of requests made to an API during a single transaction measures the efficiency and effectiveness of the API in fulfilling the needs of a specific business process.

Figure 11.11 – Product metrics for retention

In the case of APIs, the engagement is purely based on regular usage. If your customers have successfully integrated with your APIs, their usage patterns should reflect this integration. You can measure the usage patterns in a few different ways that allow you to evaluate how customers are using the APIs over a prolonged period of time, and this will help you measure their retention rate. In the following sections, you will learn about a few different retention metrics that can be insightful in evaluating your APIs.

Recurring daily, weekly, and monthly usage

Starting with raw API usage data to understand how much your customers are using your API is a great place to start. However, measuring raw API usage can be too granular to be insightful. Recurring usage is much more representative of how customers are making use of your APIs. Also, businesses usually experience usage patterns based on the day of the week, so you would need to confirm whether daily recurring usage shows a long-term pattern.

Weekly recurring usage is considered representative of the usage pattern and useful for doing long-term usage analysis. The following figure shows a chart of a customer, aggregated at a weekly grain over a 6-month period:

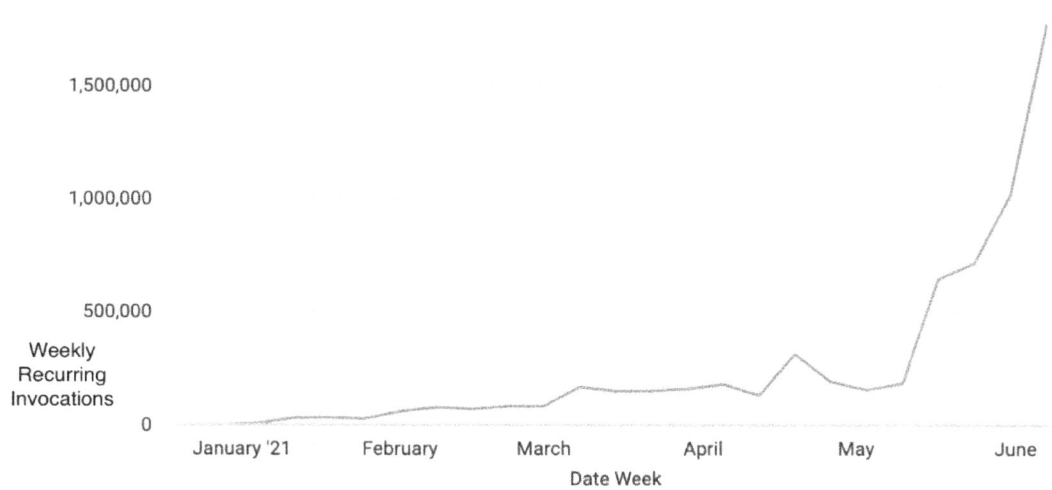

Figure 11.12 – Weekly API usage of a sample consumer over a 6-month period

When you aggregate at a weekly level, you can see the usage pick up and scale over the 6-month period. This chart helps you understand that in the case of this customer, they might have started using the APIs in January, but it took them up to the end of May to scale their usage. This shows that their integration and testing process took nearly 5 months to complete.

Customer retention

The retention rate is the number of people in a certain group who still use your product. What you think is active for your product will depend on its category. If you're using a music streaming app, you might consider playing a song each day to be an *active* day. A typical day for a payment API might be to process one credit card payment, for example.

Separating your users into groups, or *cohorts*, is essential for accurate retention measurement. This is typically done based on the date of sign-up, but when dealing with APIs, it is more useful to divide users into groups according to the date of their first successful integration – the first date of user integration and the first transaction using your API. Then, once a day, once a week, or once a month, you add up the number of unique users who came back and did whatever you think shows that they are still interested.

To measure customer retention, you can also identify the patterns of increasing usage using a rolling average of weekly usage, as we saw in the previous section. New customers might have a low rolling average of weekly usage, but the rolling average might spike as they scale and ultimately plateau once they have reached stability. When you compare the rolling average of API usage to prior weeks, you can identify both growth and churn patterns in customers, which are both very important in understanding user retention.

The only way to determine the right retention rate is to define both the first and return actions. Although it's not mandatory, the initial action is typically the first thing a new user does after installing and activating your product. Previously, we only kept tabs on first-time API callers and then recorded subsequent API calls made by those same people. It is possible to go into the specifics of those first steps or subsequent returns.

API calls per business transaction

A great way to measure the success of your API design in terms of customer value is to measure the number of API calls it takes to complete a business transaction. If your API design requires several API calls for a single transaction, this would mean the following:

- The customer takes a longer time to integrate while trying to learn about and implement all these different API calls as part of their application
- Each subsequent API call increases the chances of the transaction failing and thereby increases the risk of error for the transaction
- Multiple API calls would add a delay in the execution of the transaction

Ultimately, all these factors result in poor customer experience and/or customer value. If you set up your API endpoints so that your customers can do business with as few API calls as possible, they will be able to reach their goals.

Experience

In the previous sections, we looked at discovery, engagement, acquisition, activation, and retention metrics that trace the customer journey and measure its various aspects of it. However, as customers go through these steps in the customer journey, they are building an overall impression of the product experience. In this section, you will learn about quantitative and qualitative ways of gathering customer insights using metrics such as conversion rate, the CSAT, and the NPS.

The following diagram shows the metrics that help you measure product experience:

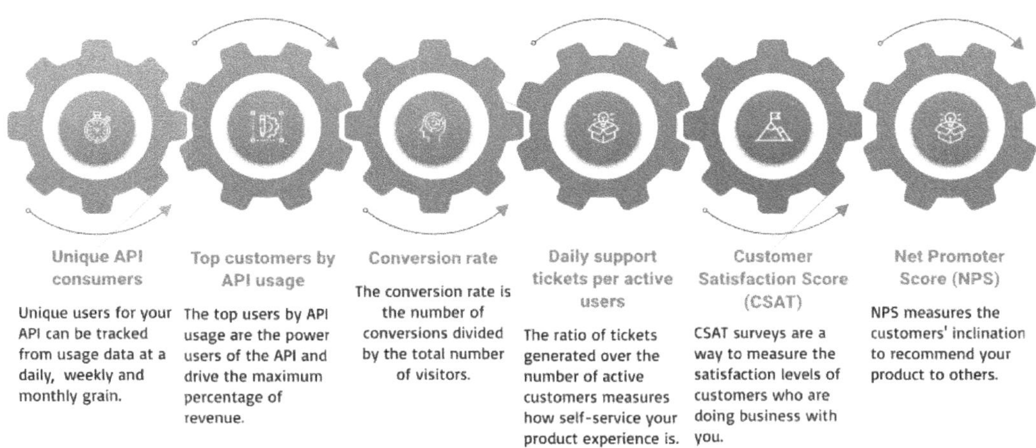

Figure 11.13 – Product metrics to measure experience

As you can see in the preceding diagram, there are six ways to measure the experience of your APIs, which we will discuss now.

Unique API consumers

APIs are primarily B2B products, and it is common to have multiple developers as part of every customer account. This might make identifying unique customers challenging in some cases. This is where your account onboarding and identity management design are key. You should work with your software architecture teams to make sure that your usage data can reliably tell you which API users are unique.

You can review account usage on a daily, weekly, and monthly basis. This would allow you to segment users based on their scale and usage. In B2B products such as APIs, you would find that a handful of customers might drive the majority of the revenue for your product. This is an important concept to understand and will be crucial to understanding your key customers.

You can analyze your customer base to understand their scale, usage patterns, and industries and whether your product has found a product-market fit in a specific customer segment. Once you develop an intimate understanding of your user, you can shape the user experience to gain more of these customers and also identify customers you can interview for more insights.

Top customers by API usage

It is common for API products to have a handful of customers that generate the majority of the revenue on your platform. As you start to look at unique API consumers and their usage, you start to observe usage patterns. Monitor your top customers by API usage to understand who your most successful customers are and what value your APIs deliver to them. You can find patterns such as scale, since not all customers will have the same scale. If you are targeting enterprise customers, you might see really high usage on a daily, weekly, and monthly basis, but if you are targeting small businesses, the scale may not be as high. Scale might also depend on where APIs are in their life cycle.

You can learn how your APIs are used most often by looking at how your top customers use them and, if possible, interviewing them. This would be insightful in identifying their journeys, in terms of how they discovered your APIs and how they made their decision to use them. Using these customers as references, you can design better user flows, develop sales strategies to acquire more resemblant audiences for your product, as well as identify upselling opportunities.

Conversion rate

The conversion rate is the number of API users that activate or reach your goal. This could be them making their first API call or sending over a thousand transactions. The goals are set on the basis of your business and how you define success for your customers. The conversion rate is a complementary metric to the TTV metric. The conversion rate is the number of customers who are able to realize the value of your product at a point in time.

Understanding this percentage helps you optimize your product onboarding and get customers integrated. The term conversion rate is often used across the organization and defined at a company level. For online shopping, the concept of the conversion rate is very closely followed, since marketing efforts are directed at getting customers to land at the site, and conversion is defined as the action of checking out and completing the transaction. In the case of APIs, defining conversion can be challenging because, as you learned in *Chapter 7*, the developer journey consists of several steps. Oftentimes, multiple developers from the same organization might be participating in evaluating a set of APIs for interaction, so from a usage data perspective, it might look like different users, but this might ultimately result in a single sale.

Sales data is the best way to track conversions for APIs because the data is usually available in a tool such as **SalesForce**, where sales teams have a defined funnel and you can track the conversion effectively.

For self-service users, you can set the rate at which they go from discovering your business to making their first purchase. This rate will allow you to track how effectively customers can discover your APIs and give you the first indication of the business impact. This metric is very useful for marketing and developer relations teams because it helps them figure out how well they are doing by breaking down the conversion rate by channel.

The formula for the conversion rate is shown here:

$$\left[\frac{Number\ of\ users\ that\ "convert"}{Number\ of\ total\ audience}\right] = Conversion\ rate$$

The conversion rate is usually reviewed on a daily, weekly, or monthly basis to ensure that there are known patterns and volumes. You can also use cohort- and campaign-based analysis to learn more about your audience and find ways to improve the customer experience.

Daily support tickets per active users

In *Chapter 6, Support Models for API Products*, you learned about the importance of the support team in your organization and how the support team is a key partners to the product manager. If your APIs have 50 active users and there are 30 customer tickets being created, then you're looking at 60% of your customers running into some kind of issue or question. In order to design a customer journey that is frictionless, it is important to provide support at various stages of the customer journey and gather insights into what those points of friction are.

Support tickets are a goldmine for qualitative insights into the customer experience. You can partner with the support team to create a tagging process and pattern to slice and dice incoming support tickets and get insight into the nature of the tickets that customers create. The first step is to create word clouds using support ticket data. This gives you a high-level sense of what the key themes are that pain your customers.

In the last section, you learned about daily, weekly, and monthly active users who come back again and again. Daily support volumes may not be a large enough sample to analyze, and the ratio of support tickets to active users should be at least weekly or monthly because this is an aggregate metric. The formula for measuring the percentage of support tickets per active account at a weekly level is shown here:

$$\left[\frac{Support\ tickets\ created\ during\ a\ week}{Active\ user\ accounts\ during\ the\ week}\right] \times 100 = Percentage\ of\ support\ tickets\ per\ active\ accounts$$

This is the metric that gives you the best idea of how well your self-service efforts are working to make it as easy as possible for customers to sign up without your help.

CSAT

If you have worked with customer success teams, you are familiar with CSAT surveys. These are a way to measure the satisfaction levels of customers who are doing business with you. They are very straightforward surveys that ask simple questions to get customer feedback and gather qualitative insights into the customer experience.

The customer satisfaction score is calculated using a five-point scale:

1. Very Unsatisfied
2. Unsatisfied
3. Neutral
4. Satisfied
5. Very Satisfied

This rating can be used to collect customer feedback for a product, service, or specific interaction. The feedback is used to compute the CSAT score using the following formula:

$$\left[\frac{\textit{The total number of 4 and 5 responses}}{\textit{Number of total responses}} \times 100\right] = \textit{Percentage of satisfied customers}$$

You can use CSAT scores as a relationship metric to evaluate customer relationships. You can also use this metric as a touchpoint metric to gather interaction-specific feedback if you are trying to evaluate a specific point in a customer journey.

Tools such as Qualtrics allow you to set up CSAT surveys easily. Depending on the size of your customer base, you can randomize the size of the audience for each survey to get a randomized sample of customer sentiment.

Net Promoter Score (NPS)

We touched upon Net Promoter Score (NPS) in *Chapter 8, Customer Expectations and Goals*, where we learned about the NPS as a metric that measures loyalty. The NPS measures a customer's inclination to recommend your product to others. Word-of-mouth marketing is one of the strongest growth levers across all products, and combined with the strong developer community of API developers who share their experiences and knowledge of APIs through various platforms, they make the NPS a crucial metric to help you understand how customers feel about your APIs. The following diagram shows what a standard NPS survey looks like:

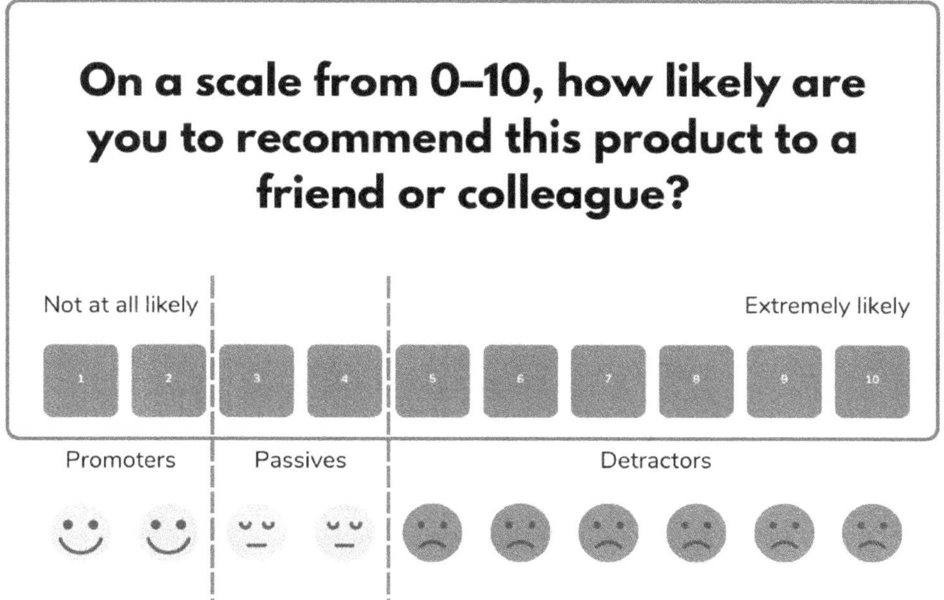

Figure 11.14 – The form fields of an NPS survey mapped to promoters, passives, and detractors

The NPS uses a scale from 0 (not likely) to 10 to estimate how likely someone is to recommend a company, product, or service to a friend or co-worker. Based on how customers answer on the 11-point scale, they are put into 1 of 3 groups: those who respond with a 9 or 10 are *promoters*, those who answer with a 7 or 8 are *passives*, and those who answer between 0 and 6 are *detractors*. The preceding diagram shows how promoters, passives, and detractors are categorized.

Passives are never used to calculate the NPS, which is calculated by simply subtracting the percentage of detractors from the percentage of promoters, as shown in the following formula:

$$Percentage\ of\ Promoters\ -\ Percentage\ of\ Detractors\ =\ NPS$$

Asking a simple question about how likely a customer is to recommend your product or service can provide invaluable insights into a customer's psyche. The NPS is a popular customer experience metric that is widely used across customer success and **user experience** (UX) organizations.

The NPS score must be run at least quarterly, if not more frequently. If you have a large user base, you can also segment or randomize the survey to make sure that customers aren't being asked for the same response too frequently.

Since the NPS measures customer sentiment after the fact, it is a lagging indicator, and you should keep that in mind as you set KPIs for your team. Usually, leading indicators that can set you on the path to success are chosen as north-star metrics for product priorities so that they can guide forward-looking product roadmaps, thus making lagging indicators such as the NPS good for measuring the results.

Summary

Building a technically sound product with a well-designed and monitored infrastructure sets your product up for success. However, it is equally important to build a smooth customer journey that guides customers to effectively discover, evaluate, and engage with your product so that they can have a great experience using it. By measuring each part of the customer life cycle, you'll be able to find ways to make a customer's journey easier and more enjoyable, which will make them more loyal to your product.

In *Chapter 10, Infrastructure Metrics*, you learned about the foundations of API analytics and how to measure your API infrastructure's performance, usage, and reliability. Building upon that foundation in this chapter, you learned how to steer customers toward success using your APIs. If your infrastructure is reliable and your product experience is enjoyable, your customers will inevitably endorse and promote your products, which will have a positive impact on your business metrics, as we will see in the next chapter.

12
Business Metrics

If you want customers to find, use, and integrate your APIs, you need to make sure they have a great product experience. If you have built a low-friction, easy-to-navigate experience to deliver high-quality, reliable, and scalable APIs, your customers will see the value; and your business metrics, such as revenue, **customer lifetime value** (**CLV**), product adoption, and so on, will start to improve.

In *Chapter 10*, you learned about infrastructural metrics and how they help evaluate the core function and quality of your APIs. In *Chapter 11, API Product Metrics*, you learned how to measure the impact of your API experience by evaluating the product metrics. In this chapter, you'll learn about the business metrics you need to set up and keep track of regularly in order to measure the business impact of your infrastructure and product development projects. You will learn about the following topics:

- Defining business metrics
- Measuring revenue
- Adoption tracking
- Churn analysis
- Optimizing for growth
- Measuring operations' efficiency

Before you dive into categories of business metrics, you will learn about the definition of business metrics in the next section.

Defining business metrics

In a small organization, the CEO of the company is often the default product manager. In medium-sized organizations, there are multiple product managers managing different aspects of the product. And in big companies, there are teams of product managers who are in charge of the different products and their features. Depending on the size of the product team you are part of, you might not have full ownership or visibility into all of the business metrics being measured and monitored across the organization, but it's important to learn about business metrics so that you can understand the priorities that are being communicated to your team.

Business metrics include measuring the financial impact of the product as well as the efficiency and effectiveness of things such as marketing, sales, and operations that don't have to do with product development. With metrics such as revenue, adoption, and churn, you can figure out how much your product is really worth. Growth metrics keep track of the marketing and sales efforts that bring in new customers. Operations metrics—such as support metrics, operations costs, and infrastructure costs—can also be used to figure out how much the costs are worth compared to the money they bring in.

Growth efforts help bring in more money for the organization, but the costs of running the business and the turnover of employees cancel out the benefits of growth. To build a profitable business, it is important to come up with a complete product strategy that will help bring in more money.

The following diagram shows the most important business metrics for APIs, including revenue, growth, adoption, churn, and operations:

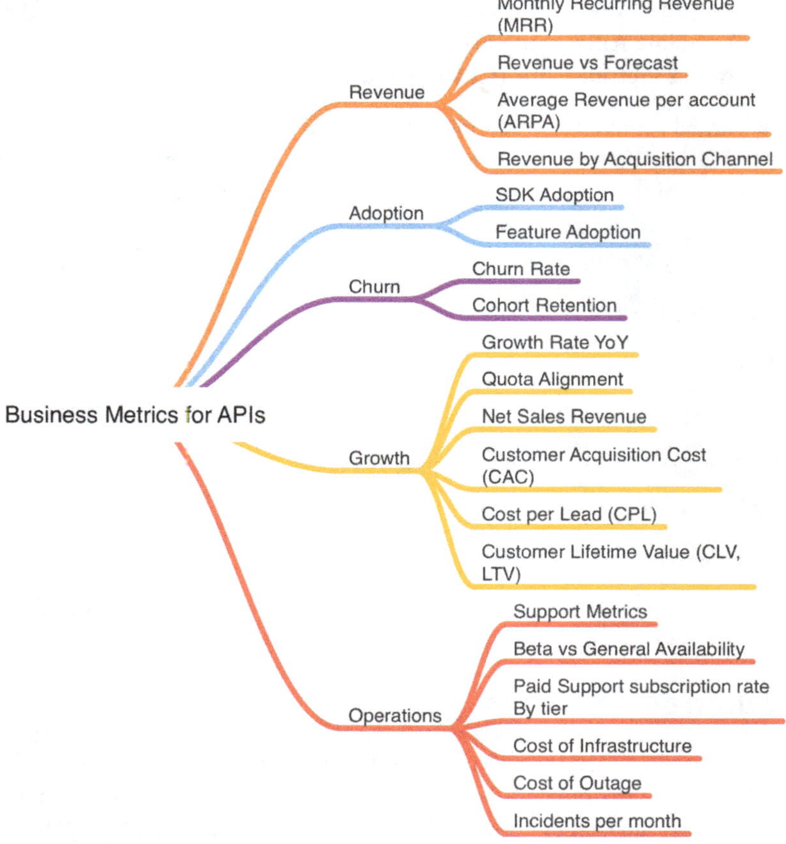

Figure 12.1 – Business metrics across revenue, adoption, churn, growth, and operations

While you may not be the primary owner of the business metrics, it is important to understand these metrics to be able to align product initiatives to these metrics so that you can track the ultimate impact of your product roadmap on the bottom line.

As shown in *Figure 12.1*, business metrics can be segmented into the following categories:

- **Revenue**: Revenue metrics measure the financial performance of a business.
- **Adoption**: Adoption metrics measure how widely a product or service is being used.
- **Churn**: Churn metrics measure the rate at which customers or users stop using a product or service.
- **Growth**: Growth metrics measure the rate at which a business is growing.
- **Operations**: Operations metrics measure the efficiency and effectiveness of a business's operations.

In the following sections, you will learn about the top metrics worth tracking across revenue measurement, adoption, churn, growth, and operations.

Measuring revenue

Revenue metrics help a business understand how much money it generates from its products or services. These metrics can include total revenue, growth, average transaction value, and CLV.

Tracking revenue is essential for several reasons. First and foremost, revenue is the primary driver of a business's profitability. By tracking revenue metrics, a company can understand how well it performs financially and make informed decisions about allocating its resources.

Tracking revenue can help a business figure out how well it is doing financially and also spot trends and patterns in how customers act. For example, a company might see that a certain product or service is making a lot of money, which shows that there is a lot of demand for it. A business can decide which products or services to invest in or stop offering by looking at its sales data.

Tracking revenue metrics is an integral part of running a business because it helps a company understand its financial performance, find trends and patterns in how customers act, and make intelligent decisions about where to put resources.

The following screenshot shows the key revenue metrics that you should track for your API business:

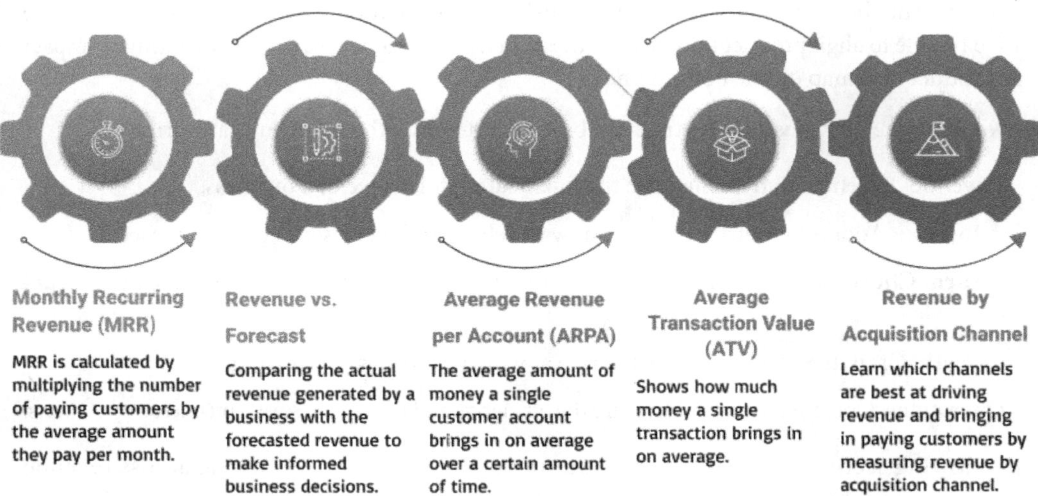

Monthly Recurring Revenue (MRR)
MRR is calculated by multiplying the number of paying customers by the average amount they pay per month.
Revenue vs. Forecast
Comparing the actual revenue generated by a business with the forecasted revenue to make informed business decisions.
Average Revenue per Account (ARPA)
The average amount of money a single customer account brings in on average over a certain amount of time.
Average Transaction Value (ATV)
Shows how much money a single transaction brings in on average.
Revenue by Acquisition Channel
Learn which channels are best at driving revenue and bringing in paying customers by measuring revenue by acquisition channel.

Figure 12.2 – Revenue metrics for APIs

In the following subsections, you will learn about key revenue metrics such as **Monthly Recurring Revenue** (**MRR**), **Average Revenue per Account** (**ARPA**), and so on.

MRR

MRR is a way to figure out how much predictable, recurring income a business can expect to get every month. It is often used to track the performance of subscription-based businesses, as it provides a clear and consistent way to measure the growth of a company's revenue over time. MRR also helps in planning immediate business needs, such as operations and hiring.

MRR is calculated by multiplying the number of paying customers by the average amount they pay per month. For example, if a business has 100 paying customers who each pay $50 per month, its MRR would be $5,000 (100 x $50).

Measuring MRR is important for API products because it allows a business to understand the long-term revenue potential of its API offerings. By tracking MRR over time, a business can see how well its API products are performing and make data-informed decisions about how to allocate resources and improve its products.

MRR is also a key metric for subscription-based businesses because it gives a consistent way to measure how a company's revenue grows over time. Even if your pricing is not subscription-based, you should still be measuring MRR so that you can correlate it to customer loyalty and CLV. By tracking MRR, a business can see how well it is retaining and acquiring customers, and make informed decisions about its pricing and marketing strategy.

Overall, measuring MRR is important for API products because it allows a business to understand the long-term revenue potential of its API offerings, track the growth of its revenue over time, and inform its product roadmap, sales, and marketing strategies.

Revenue versus forecast

Revenue is how much money a business or product makes by selling goods or services. When it comes to APIs, how much money you make depends on how you price API usage. Businesses use revenue as a key metric to measure their performance and success.

Forecast analysis looks at a business's past performance and market trends to predict how much money it will make and how much it will spend in the future. This allows businesses to make informed decisions about their operations and make adjustments as needed to meet their financial goals.

Revenue versus forecast analysis involves comparing the actual revenue generated by a business with the forecasted revenue. This can help a business figure out whether there are any differences or trends between the two and make any needed changes to how they run their business or make predictions. By regularly performing revenue versus forecast analysis, businesses can stay on track and make informed decisions to drive growth and profitability.

ARPA

ARPA is a metric that shows how much money a single customer account brings in on average over a certain amount of time. It is commonly used by **software-as-a-service** (**SaaS**) companies to track the financial performance of their products and services.

To figure out ARPA, divide the total amount of money made from all customer accounts during a certain time period by the number of customer accounts. The formula for ARPA is shown here:

$$\frac{Total\ Revenue}{Number\ of\ Accounts} = Average\ Revenue\ per\ Account\ (ARPA)$$

If a SaaS company earns $100,000 per month from 1,000 customer accounts, ARPA is $100 ($100,000/1,000).

ARPA is a useful metric for SaaS companies to track because it helps them figure out how much each customer is worth and find ways to make more money by upselling or cross-selling. It can also help a SaaS company understand how its products and services perform as a whole and find places to improve.

ARPA can be a very important metric to track for a SaaS API product because it helps the company understand how valuable its API is to customers and find growth opportunities. By keeping track of ARPA on a regular basis, a SaaS API company can make smart decisions about its prices and products in order to increase sales and profits.

Average transaction value (ATV)

ATV is a metric that shows how much money a single transaction brings in on average. ATV is commonly used in the context of e-commerce and other businesses that sell products or services on a per-transaction basis.

In the case of APIs, *average transaction value* may refer to the average revenue generated from a single API request or transaction. For example, if a company charges customers for using its API on a per-request basis, it could track the average revenue generated from each API request to measure the overall performance of its API offering.

ATV can be a useful metric for API companies to track, as it can help them understand the customer segments your APIs are most successful with. ATV can not only inform revenue and profitability, but it can also inform your pricing strategy and help identify the product market fit for your APIs.

It's important to remember that ATV can change depending on how an API company runs its business and sets its prices. Some API companies may charge customers on a subscription basis, while others may charge on a per-request or per-transaction basis. In any case, tracking ATV can help an API company understand the value of its API offering to customers and make informed decisions to drive revenue and profitability.

Revenue by acquisition channel

API companies can learn which channels are best at driving revenue and bringing in paying customers by measuring revenue by acquisition channel. By keeping track of revenue by acquisition channel, an API company can figure out which channels bring in the most money and put their resources there to make their marketing and sales work as effectively as possible.

Tracking revenue by channel of acquisition can also help you find any differences or trends between channels and make any changes to their marketing and sales strategies that are needed. For example, if you see that one channel is consistently underperforming compared to others, you may decide to put fewer resources into that channel and focus on others that are better at driving revenue.

Adoption tracking

Adoption metrics are measurements that show how much customers use or adopt a product or service. In the context of APIs, adoption metrics can help a company understand how its API is being used and how effective it is at meeting the needs of customers.

The following screenshot shows the two key adoption metrics that are important to track for business stakeholders:

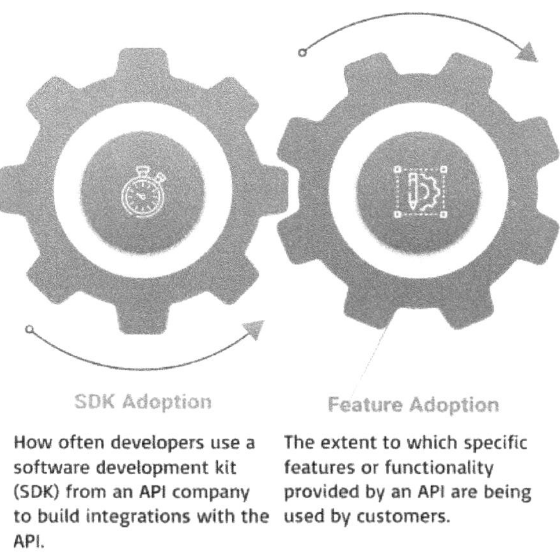
SDK Adoption
How often developers use a software development kit (SDK) from an API company to build integrations with the API.
Feature Adoption
The extent to which specific features or functionality provided by an API are being used by customers.

Figure 12.3 – Key adoption metrics for business stakeholders

By keeping track of adoption metrics on a regular basis, an API company can see how its API is used and how well it works. This lets the company make smart decisions to drive growth and improve the product. You will learn more about measuring **software development kit** (**SDK**) and feature adoption in the sections that follow.

SDK adoption

SDK adoption metrics track how often developers use an SDK from an API company to build integrations with the API. Some examples of SDK adoption metrics for APIs include:

- **Number of SDK downloads**: This measures the number of times the SDK has been downloaded by developers. It can help an API company understand the popularity of its SDK and identify opportunities for growth.
- **Number of unique developers**: This measures the number of unique developers who have downloaded the SDK. It can help an API company understand the reach of its SDK and identify opportunities for growth.
- **Number of integrations built using the SDK**: This measures the number of third-party systems or applications that have been built using the SDK. It can help an API company understand how its SDK is being used in different contexts and identify opportunities for growth.

SDKs are instrumental in reducing integration time for customers, and by tracking SDK version adoption you can track which SDKs are in use and which ones can be deprecated so that you can improve the maintainability of your SDKs. You can also track feature adoption with respect to different SDKs. You will learn more about feature adoption in the next section.

Feature adoption

Feature adoption metrics measure the extent to which specific features or functionality provided by an API are being used by customers. Some examples of feature adoption metrics for APIs include:

- **Number of feature requests**: This measures the number of times a specific feature has been requested by customers. It can help an API company understand the demand for a particular feature and prioritize development efforts accordingly.
- **Feature usage rate**: This measures the percentage of customers who are using a specific feature. It can help an API company understand the value of a particular feature to customers and identify any issues that may be preventing wider adoption.

Churn analysis

Churn, also known as customer churn or user churn, refers to the rate at which customers or users stop using a product or service. It is a way to measure how well a product or service keeps its customers or users. It is usually given as a percentage of the total number of customers or users who have stopped using the product or service over a certain time period.

The following screenshot shows the two key metrics you will learn about in this section:

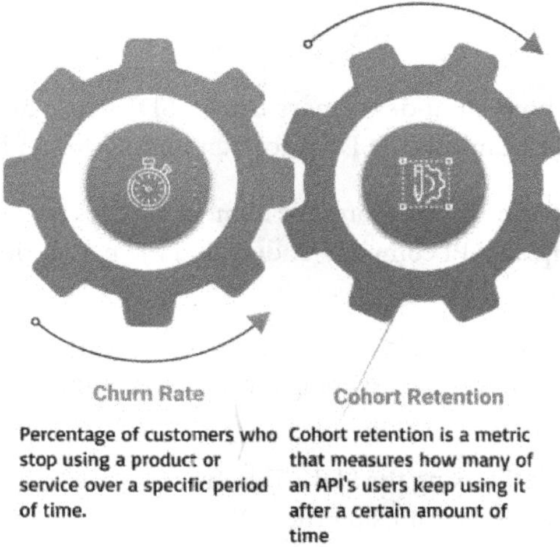
Churn Rate
Percentage of customers who stop using a product or service over a specific period of time.
Cohort Retention
Cohort retention is a metric that measures how many of an API's users keep using it after a certain amount of time

Figure 12.4 – Churn and retention metrics for business stakeholders

Churn is an important metric for businesses to track, as it can have a significant impact on revenue and growth. High churn rates can indicate that a business is not meeting the needs of its customers or that there are issues with the product or service that are causing customers to leave. By keeping track of churn, a business can find problems and take steps to keep customers, ultimately improving customer retention.

Churn rate

The churn rate is the percentage of customers who stop using a product or service over a specific period of time. In the context of API products, the churn rate can help a company figure out how well it is keeping customers and find any problems that may be causing customers to stop using the API.

To calculate the churn rate, you would divide the number of customers who stopped using your API over a specific period of time by the total number of customers at the beginning of that period. For example, if you had 100 customers at the beginning of a month and 10 of those customers stopped using the API by the end of the month, the churn rate would be 10% (10/100). The formula for calculating the churn rate is shown here:

$$\left[\frac{Number\ of\ customers\ who\ churned\ in\ a\ specific\ period\ of\ time}{Total\ number\ of\ customers\ at\ the\ beginning\ of\ that\ period}\right] = Churn\ Rate$$

There are three key churn rate metrics that you can measure for your APIs:

- **Customer churn rate**: This measures the percentage of customers who stop using the API over a specific period of time. It can help you understand the overall retention of its customers and identify any trends or changes in churn over time.
- **Revenue churn rate**: This measures the percentage of revenue that is lost due to customers stopping their use of the API. It can help you understand the financial impact of churn and identify any issues that may be causing customers to stop using the API.
- **Feature churn rate**: This measures the percentage of customers who stop using a specific feature of the API. It can help you understand the value of a particular feature to customers and identify any issues that may be causing customers to stop using it.

Churn rate can be an important metric for API companies to track, as it can help them understand the retention of their customers and identify any issues that may be causing customers to stop using the API.

Cohort retention

Cohort retention is a complementary metric to churn rate. Cohort retention is a metric that measures how many of an API's users keep using it after a certain amount of time. It is usually measured monthly or quarterly.

The formula for calculating the retention rate is shown here:

$$\left[\frac{Number\ of\ customers\ who\ continued\ usage\ in\ a\ specific\ period\ of\ time}{Total\ number\ of\ customers\ at\ the\ beginning\ of\ that\ period}\right] = Retention\ Rate$$

Some common cohort retention metrics for APIs include:

- **Monthly Active Users** (**MAUs**): This metric measures the number of unique users who made at least one API call in a given month
- **Monthly Recurring Users** (**MRUs**): This metric measures the number of unique users who made at least 1 API call in each of the past 3 months
- **Monthly Retention Rate**: This metric measures the percentage of users who continue to use the API in a given month
- **Quarterly Retention Rate**: This metric measures the percentage of users who continue to use the API over a 3-month period

It's important to remember that these metrics are only useful when compared to previous periods or to industry benchmarks. This is because they can help you find trends in API usage and make decisions about how to improve retention that are based on accurate information.

Optimizing for growth

Growth metrics help evaluate an API's success and growth. These metrics can help identify areas of strength and areas for improvement in the API and can be used to set goals and track progress over time. Growth metrics focus on converting new and potential customers into existing customers. Sales and marketing teams are vital growth drivers at this part of the customer journey. The developer relations team also helps get more people to use your APIs because the content and community they build make it easier for developers to use them. Sales teams often work with larger customers where there are multiple decision-makers, and early customer targeting is managed through tools such as Salesforce, where sales teams track their efforts.

In this early phase of the customer journey, you have a few key indicators of growth that can help you measure your growth efforts. As you shape your APIs' marketing and sales strategy, measuring product and infrastructure metrics is only part of the puzzle. You can measure the sales and marketing efforts in a way that allows you to calculate the targeting, response, and acquisition costs to ensure that the growth is in the desired direction. In this section, you will learn about metrics such as lead response, growth rate, quota alignment, and so on that help bring the sales and marketing teams into alignment with the product growth efforts.

The following screenshot shows the key growth metrics that you will learn about in this section:

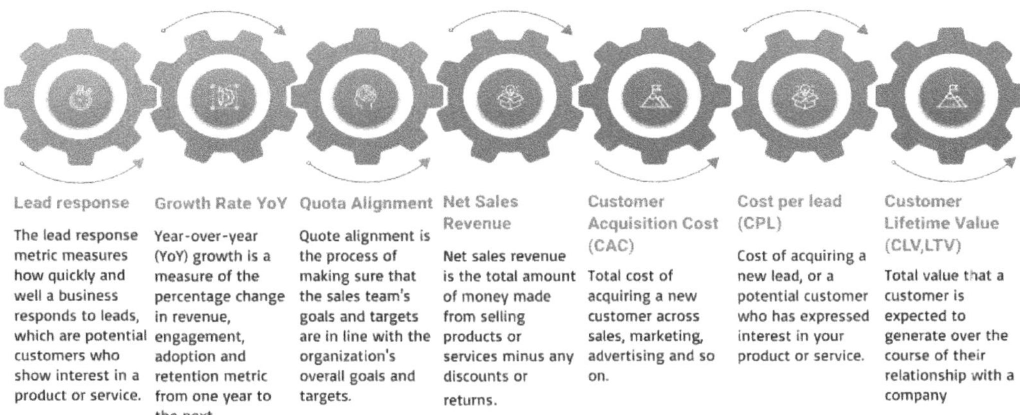

Lead response
The lead response metric measures how quickly and well a business responds to leads, which are potential customers who show interest in a product or service.
Growth Rate YoY
Year-over-year (YoY) growth is a measure of the percentage change in revenue, engagement, adoption and retention metric from one year to the next
Quota Alignment
Quote alignment is the process of making sure that the sales team's goals and targets are in line with the organization's overall goals and targets.
Net Sales Revenue
Net sales revenue is the total amount of money made from selling products or services minus any discounts or returns.
Customer Acquisition Cost (CAC)
Total cost of acquiring a new customer across sales, marketing, advertising and so on.
Cost per lead (CPL)
Cost of acquiring a new lead, or a potential customer who has expressed interest in your product or service.
Customer Lifetime Value (CLV,LTV)
Total value that a customer is expected to generate over the course of their relationship with a company

Figure 12.5 – Growth metrics for APIs tracked by business stakeholders

In the following sections, you will learn in depth about the growth metrics such as lead response, growth rate **Year-Over-Year** (**YoY**), **Customer Acquisition Cost** (**CAC**), and so on tracked by sales, marketing, and other business stakeholders.

Lead response

A lead is a potential customer who has expressed interest in a product or service. Leads can come from a variety of sources, such as online forms, social media, or in-person events. A qualified lead is a lead that has been evaluated and deemed likely to convert into a customer. This evaluation process, also called lead qualification, usually involves gathering information about the lead's needs, budget, and decision-making process to see whether they are a good fit for the product or service being offered.

The lead response metric measures how quickly and well a business responds to leads, which are potential customers who show interest in a product or service. This metric is important because it can have a significant impact on the success of the business.

A few different ways you can measure lead response time include:

- **First response time**: This metric measures the time it takes for a business to respond to a lead for the first time. A shorter response time can be more effective in engaging and converting leads.
- **Time to contact**: This metric measures the time it takes for a business to make contact with a lead, either through a phone call or an in-person meeting.

- **Time to qualify**: This metric measures the time it takes for a business to determine whether a lead is qualified or likely to convert into a customer.

By keeping track of these metrics, businesses can find any slowdowns or inefficiencies in how they respond to leads and make changes to improve their chances of turning leads into customers. Some SaaS APIs may come with tools or features that help businesses automate and optimize their lead response process. This can improve the lead response metric and make it more likely that leads will turn into customers.

Growth rate YoY

Conversion refers to the process of turning a lead into a customer. This can mean doing things such as following up with the lead, giving the lead more information about the product or service, and negotiating the terms of a sale. Growth is a factor in increasing the conversion of new customers in a controllable and repeatable way.

Businesses often use email marketing, social media marketing, and personalized sales outreach, among other things, to nurture and engage leads so that they are more likely to become customers. By keeping track of metrics such as the number of leads generated, the percentage of qualified leads, and the conversion rate (the percentage of leads that are successfully turned into customers), businesses can measure the effectiveness of their sales efforts and find areas to improve.

To measure the growth rate YoY for an API, you will need to track certain metrics over time and compare the results from one year to the next. Some common metrics that can be used to measure the growth rate of an API include:

- **Usage**: This metric measures the number of API calls or requests made to the API. To measure the YoY growth rate, you will need to track the number of API calls for each year and calculate the percentage increase or decrease from one year to the next.
- **Retention**: This metric measures the percentage of users who continue to use the API over time. To measure the YoY growth rate, you will need to track the retention rate for each year and calculate the percentage increase or decrease from one year to the next.
- **Engagement**: This metric measures the level of interaction and activity with the API, such as the number of unique users or the average time spent using the API. To measure the YoY growth rate, you will need to track these metrics for each year and calculate the percentage increase or decrease from one year to the next.
- **Revenue**: For APIs that charge a fee for usage, revenue is an important growth metric. To measure the YoY growth rate, you will need to track the revenue generated by the API for each year and calculate the percentage increase or decrease from one year to the next.

YoY growth is a measure of the percentage change in a metric from one year to the next. The formula for calculating YoY growth is shown here:

$$\left[\frac{(Current\ year\ value\ -\ Previous\ year\ value)}{Previous\ year\ value}\right] = YoY\ Growth$$

For example, if a company's revenues in the current year are $100,000 and its revenues in the previous year were $80,000, the YoY growth in sales would be:

YoY growth = ($100,000 - $80,000) / $80,000 = 25%

When executives look at metrics, they often look at growth YoY metrics, which are aggregated metrics built on top of product and infrastructure metrics. By tracking these metrics and calculating the YoY growth rate, you can get a sense of how the API is performing and whether it is growing over time. This can help you identify areas of strength and areas for improvement and set goals for future growth.

Quota alignment

In the context of sales teams, a quota refers to a specific sales target that is assigned to each salesperson or sales team. This target represents the amount of revenue or the number of sales that the salesperson or team is expected to achieve within a certain period of time, such as a month or a quarter. Quotas can be set for different levels of sales, such as per product, per team, per territory, per region, and so on. They can also be set for different timeframes, such as monthly, quarterly, or annually.

Quotas are commonly used by sales managers to set expectations for their teams and to measure the performance of individual salespeople. Quotas are also used to align the sales team's efforts with the overall goals and objectives of the company.

Quota alignment is the process of making sure that the sales team's goals and targets are in line with the organization's overall goals and targets. This can be especially important for an API sales team since APIs can be a key part of a company's product or service.

In order for an API sales team to be effective, it is important that they have has a clear understanding of the goals of the organization and how their efforts contribute to achieving those goals. This may involve setting targets and metrics for the team to meet, such as the number of API integrations or the value of API sales. It may also involve establishing processes and systems for tracking progress and measuring success.

Effective quota alignment can help make sure that the API sales teams are working toward the right goals and that their efforts are in line with the organization's overall strategy. It can also help motivate and engage team members because they know exactly what they need to do and how their work affects the company's success.

There is no set formula for quota alignment because the way sales goals and quotas are matched up depends on the needs and goals of the organization. However, some general principles can be followed when aligning sales quotas with organizational goals:

1. Identify the overall goals and objectives of the organization. This could include financial targets, market share goals, or other objectives.

2. Determine how the sales team can contribute to achieving those goals. This might involve setting targets for the number of API integrations, the value of API sales, or other metrics that align with the organization's goals.
3. Communicate the goals and targets to the sales team. It is important that all team members have a clear understanding of what is expected of them and how their efforts contribute to the success of the organization.
4. Establish processes and systems for tracking progress and measuring success. This might involve setting up regular check-ins or using software to track progress and measure performance.
5. Continuously review and adjust quotas as needed. As the organization's goals and priorities change, it may be necessary to adjust quotas to ensure that they remain aligned with the overall strategy of the organization.

It is also important to consider the needs and capabilities of the sales team when setting quotas. Quotas that are too high may be unrealistic and demotivating, while quotas that are too low may not challenge the team and may not contribute to the organization's goals.

Net sales revenue

Net sales revenue is the total amount of money made from selling products or services minus any discounts or returns. In the context of API sales, *net sales revenue* would refer to the total value of API sales, minus any discounts or returns.

Most of the time, licensing or selling access to an API is how money is made from API sales. Companies may offer APIs so that outside developers or organizations can connect their products or services to the company's systems or platform. API sales revenue can be an important source of revenue for companies that rely on API sales as part of their business model.

To calculate net sales revenue in the context of API sales, you would first need to determine the total value of API sales. This could include the value of API licenses or subscriptions as well as any additional revenue generated through the use of the API (such as fees for API calls or usage). Then, you would need to subtract any discounts or returns that were applied to API sales. The resulting number would be the net sales revenue for API sales.

Net sales revenue is an important metric for companies to track because it shows how well the company is doing overall and how well its finances are doing. It can be used to track the success of API sales efforts and set sales strategies.

CAC

CAC is the total cost of acquiring a new customer, and it can be an important metric for businesses to track. When talking about an API product, CAC means the cost of getting a new customer who uses the API.

To figure out CAC for an API product, you would have to add up all the costs of getting a new customer, such as marketing and sales costs. Some specific items to consider when calculating CAC for an API product might include:

- **Advertising costs**: This could include costs for paid advertising, such as Google AdWords or social media advertising.
- **Sales and marketing salaries**: This would include the salaries of any sales or marketing team members who are involved in acquiring new customers for the API product.
- **Lead generation and qualification costs**: This could include costs for lead generation efforts, such as website development or content marketing, as well as the costs of qualifying leads before they become customers.
- **Community building costs**: Developer relations teams are a key driver of community building around your APIs. These efforts often require a team that is involved with content generation, organizing events and workshops, as well as sponsoring conferences, among other activities.
- **Customer onboarding costs**: This could include any costs associated with onboarding new customers, such as training or support expenses.

To calculate CAC, you would add up all of these costs and divide them by the number of new customers acquired over a given period of time (such as a month or a year). This will give you the average CAC for the API product.

It's important to keep track of CAC over time and compare it to other metrics, such as revenue or CLV, to figure out how well your customer acquisition efforts are working overall. CAC can also be used to find ways to save money or better use the resources you have.

Cost per lead (CPL)

CPL is a metric that measures the cost of acquiring a new lead, or a potential customer who has expressed interest in your product or service. In the context of APIs, CPL would refer to the cost of acquiring a new lead for an API product.

To calculate CPL for an API product, you would need to consider all of the costs associated with acquiring a new lead, including advertising, marketing, and sales expenses. To figure out CPL, you would add up all of these costs for advertising, lead generation and qualification, sales, and marketing, then divide by the number of new leads you got in a certain amount of time (such as a month or a year). This will give you the average CPL for the API product.

CPL is an important metric for companies to keep track of because it can show how well lead generation efforts are working and help find ways to save money or improve performance. It is important to compare CPL to other metrics, such as conversion rate and CAC, to get a complete picture of the effectiveness of your lead generation efforts.

CPL and CAC are both metrics that measure the cost of acquiring new customers or leads. However, there is a key difference between the two metrics: CPL measures the cost of acquiring a new lead, while CAC measures the cost of acquiring a new customer.

A lead is a potential customer who has expressed interest in a product or service but has not yet made a purchase. CPL measures the cost of acquiring a new lead or generating interest in a product or service. This might include the cost of advertising or marketing efforts that are aimed at generating leads, as well as the cost of qualifying leads before they become customers.

On the other hand, CAC measures the total cost of acquiring a new customer, or someone who has actually made a purchase. This includes not only the costs of acquiring a lead but also the additional costs associated with converting a lead into a paying customer, such as onboarding expenses or customer support costs.

Both CPL and CAC are important metrics for businesses to track, as they provide insight into the efficiency and effectiveness of customer acquisition efforts. However, it is important to understand the difference between the two metrics and to consider them in the context of the overall performance of the business. CAC, in particular, is an important metric for evaluating the overall profitability of a business, as it takes into account the total cost of acquiring a customer, including the costs of converting leads into paying customers.

CLV/LTV

CLV or LTV is a measure of the total value that a customer is expected to generate over the course of their relationship with a company. In the context of APIs, CLV would refer to the total value that a customer is expected to generate through the use of an API product.

To calculate CLV for an API product, you would need to consider a number of variables, including:

- **The value of each API transaction**: This could include the cost of API licenses or subscriptions, as well as any additional revenue generated through the use of the API (such as fees for API calls or usage).
- **The frequency of API transactions**: This would take into account how often the customer is expected to make API transactions over the course of their relationship with the company.
- **The length of the customer relationship**: This would consider the expected duration of the customer relationship, which could be influenced by factors such as the customer's satisfaction with the API product and the overall stability of the market.

To calculate CLV, you would need to multiply the value of each API transaction by the frequency of API transactions, and then multiply that number by the length of the customer relationship. This would give you an estimate of the total value that a customer is expected to generate over the course of their relationship with the company.

The formula for calculating CLV/LTV is shown here:

$$(average\ value\ of\ a\ transaction) \times (number\ of\ transactions\ per\ year) \times (average\ customer\ lifespan) = Customer\ Lifetime\ Value\ (CLV/LTV)$$

For example, if the average value of an API transaction is $0.2, the customer is expected to make 10 million API transactions per year, and the average customer lifespan is 5 years, the CLV would be:

CLV = ($0.2) x (10 million) x (5) = $10 million

It is important to note that the values used in the CLV formula will depend on the specific business and the nature of the customer relationship. The average value of a transaction can change depending on the type of API product being sold and the pricing model being used. The number of transactions per year will depend on how often customers use the API product, and the average customer life expectancy will depend on how long customers are expected to stay with the company.

CLV is an important metric for companies to track, as it provides insight into the overall value of a customer relationship and can help to inform business decisions related to customer acquisition, retention, and monetization. In the context of APIs, CLV can be a useful metric for finding ways to make API transactions more valuable or to retain customers for longer.

Measuring operations' efficiency

Operational metrics are important for any business, but they are particularly important for API businesses. This is because API businesses depend on the cloud to deliver their software, and operational metrics can help make sure that the software is always and reliably delivered.

Some specific ways in which operational metrics can be important for a SaaS business include:

- **Identifying problems and bottlenecks**: Operational metrics can help to identify problems or bottlenecks in the delivery of the software, such as slow response times or high error rates. This can help to identify areas that need improvement and optimize the delivery of the software.
- **Monitoring performance**: Operational metrics can help to monitor the overall performance of the software, including metrics such as uptime and availability. This can help to ensure that the software is performing as expected and meeting the needs of users.
- **Providing insights into customer behavior**: Operational metrics, such as API usage and adoption rates, can give information about how customers use the software. This can help to inform product development and marketing efforts.
- **Improving customer satisfaction**: By monitoring and improving operational metrics, a SaaS business can help to improve customer satisfaction by ensuring that the software is being delivered consistently and reliably.

Overall, operational metrics are important tools for any SaaS business, as they can help to ensure that the software is being delivered effectively and efficiently and that the business is meeting the needs of its customers. The following screenshot shows the six key operations metrics that are notable for business decision-making processes:

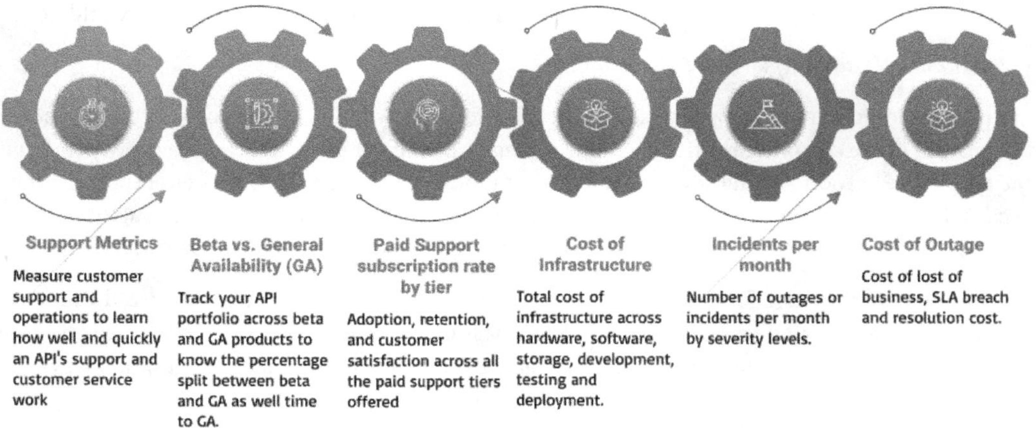

Support Metrics
Measure customer support and operations to learn how well and quickly an API's support and customer service work
Beta vs. General Availability (GA)
Track your API portfolio across beta and GA products to know the percentage split between beta and GA as well time to GA.
Paid Support subscription rate by tier
Adoption, retention, and customer satisfaction across all the paid support tiers offered
Cost of Infrastructure
Total cost of infrastructure across hardware, software, storage, development, testing and deployment.
Incidents per month
Number of outages or incidents per month by severity levels.
Cost of Outage
Cost of lost of business, SLA breach and resolution cost.

Figure 12.6 – Operations metrics for APIs

In the following sections, you will learn in depth about operations metrics that are tracked at a business level, such as support metrics, cost of infrastructure, incidents per month, and so on.

Support metrics

We looked at support models and metrics for APIs in depth in *Chapter 6*. Metrics for support and customer service are important for figuring out how well and quickly an API's support and customer service work. Support and customer service metrics are kept track of by executives so that they can make strategic decisions about hiring staff or setting up operations in new places. The key support metrics that have a direct business impact include:

- **Ticket volume**: The number of support tickets or customer service requests received by the API team.
- **Ticket resolution time**: The time it takes for the API team to resolve support tickets or customer service requests.
- **Customer satisfaction**: The level of satisfaction expressed by API users in response to support or customer service inquiries. This can be measured through surveys or other methods.

- **First response time**: The time it takes for the API team to respond to a support ticket or customer service request.
- **Average handle time**: The average amount of time it takes for the API team to complete a support ticket or customer service request.
- **Escalation rate**: The percentage of support tickets or customer service requests that need to be escalated to higher levels of support.

By keeping track of these metrics, you can measure how effective and efficient its support and customer service efforts are, find places to improve, and make sure API users get the help they need when they need it.

Beta versus General Availability (GA)

The difference between beta APIs and APIs that are available to everyone shows where the API is in its development and how ready it is to be used. Beta APIs are still being tested and worked on, while GA APIs are stable and ready for the public to use.

Most organizations have hundreds—if not thousands—of APIs. When you set clear standards for APIs to move from beta to GA, you can track the following metrics across your API portfolio:

- **Percentage of APIs in beta**: The percentage of APIs in your API portfolio across internal, partner, and public APIs that are in beta
- **Percentage of APIs in GA**: The percentage of APIs in your API portfolio across internal, partner, and public APIs that are generally available
- **Time to GA**: The time it takes for your team to take API offerings from production to GA status

An API that is in GA is an API that has completed the testing and development phase and is available for general use by the public. APIs with GA are typically fully supported and intended for widespread use by a diverse range of users and are frequently accompanied by documentation and other resources to assist users in understanding how to use the API. Quality is shown by making sure that as many of your APIs as possible meet this standard. Tracking how long it takes for your teams to deliver GA APIs can help you evaluate developer productivity and plan roadmaps more effectively.

Paid support subscription rate by tier

Most SaaS companies offer paid support options to customers. Paid support lets customers with complicated integrations and a lot of dependencies get help right away if something goes wrong. Support is often thought of as a cost center, but for enterprise customers, paid support is a must-have, and the quality of support is what keeps customers coming back.

Keeping track of adoption, retention, and customer satisfaction across all the paid support tiers your company offers can make sure that the best experience goes to the most valuable customers.

Cost of infrastructure

Building APIs is a complex task that requires both time and money. When you hire teams of developers, security experts, marketing experts, sales experts, and so on, you have to pay for them. But there is also a cost associated with the underlying infrastructure that hosts your APIs.

The cost of infrastructure for APIs can be calculated in a number of different ways, depending on the specific requirements of the API and the resources that are needed to support it. Some factors that may be taken into account when calculating the cost of infrastructure for APIs include:

- **Hardware and software costs**: The cost of hardware and software needed to support the API, such as servers, storage, and network infrastructure
- **Maintenance and support costs**: The cost of maintaining and supporting the API infrastructure, including the cost of technical support, updates, and repairs
- **Bandwidth and storage costs**: The cost of bandwidth and storage needed to support the API, which can vary depending on the volume of traffic and data being processed
- **Development and testing costs**: The cost of developing and testing the API, including the cost of hiring developers, building and maintaining the API, and conducting testing and quality assurance
- **Deployment and hosting costs**: The cost of deploying and hosting the API, including the cost of hosting servers and other infrastructure in a cloud or on-premises environment

Overall, the cost of API infrastructure will depend on a number of factors, such as the API's complexity, the resources needed to support it, and the needs of the organization that is putting it in place.

Incidents per month

Service-level agreements (**SLAs**) are contracts that every company that makes APIs makes with its customers. These contracts say how the customer's business losses will be covered in the event of an incident. Incidents could be caused by anything, from a problem with the infrastructure to a bug that came with a new feature. As an API provider, you should try to make sure that your APIs are as reliable as possible. However, incidents will happen, so there must be ways to measure the impact of an incident, communicate with the customer to set their expectations, and evaluate the business impact of an incident.

In *Chapter 10*, you learned how to measure the reliability of your APIs using the **mean time to failure** (**MTTF**), **mean time to repair** (**MTTR**), **mean time between failures** (**MTBR**), and **probability of failure on demand** (**POFOD**) metrics. Even though these metrics are very specific and are usually tracked from an infrastructure point of view, the business impact of reliability metrics is focused on measuring incidents per month.

Twilio is an example of a company that puts out clear SLAs based on uptime thresholds. If something happens that affects the uptime of applicable APIs and breaks the 99.99% uptime metric, Twilio gives its customers 10% of a service credit. This results in a direct impact on revenue. This is why it's important to keep track of how many incidents happen each month because each one costs money and hurts customer trust.

Cost of outage

Leading directly from the incidents per month metric we saw previously is the cost of the outage. Each time there is an outage, there is a loss of business for you and for your customers. This loss is not only monetary, but it also represents a loss of trust from your customers.

Because outages cost money and have bad effects on business, engineering teams keep a close eye on reliability metrics. The length and severity of the outage are two of the most important factors in figuring out how much it will cost. If there is a less severe outage that only lasts a small amount of time, you might not have a very big revenue impact. But if an outage is very bad and takes engineering teams a long time to find the problem and fix it, the outage could last longer and have a much bigger effect on the business.

From a business perspective, it's important to monitor the cost of outages to make sure that it is within an acceptable range. When engineering teams are making APIs, the requirements can be clearly communicated as a set of goals, and APIs that don't meet these standards should not be shown to customers as mature, ready-to-use products. This acceptable range should be established by the leaders in alignment with all the stakeholders so that customer expectations can be set accordingly.

Summary

Building good APIs starts with making them easy to use and reliable, scalable, and robust. Packaging your APIs with a great API experience and helping customers discover your APIs and effectively evaluate, integrate, test, and scale their usage helps you onboard customers successfully. If your customers are able to use your APIs effectively and build successful applications using them, growth patterns will show up in your business results.

Building an analytics strategy in steps, starting with infrastructure metrics as the base, then moving on to product metrics, and then connecting them to business metrics, makes sure that everyone working on the product has the same goals and priorities. Now that you are familiar with numerous metrics across various dimensions of infrastructure, product, and business, in the upcoming chapters, you will learn about evaluating these metrics against each other and clustering them to create various leading and lagging metrics, **input and output** (**I/O**) metrics, and so on to identify, prioritize, and interpret these metrics to guide your decision-making across all stakeholders.

Part 4: Setting a Cohesive Analytics Strategy

In *Part 3*, you learned about the vast set of metrics you can use to measure infrastructure, product, and business initiatives. As you begin to capture the data and start to analyze the metrics, you will need to be able to interpret these metrics, identify the right **key performance indicators** (**KPIs**) for your product, and make decisions based on these metrics. In this part, you will learn about how you can stitch together the insights from the various metrics you have, evaluate and contrast each of them to validate or invalidate your findings, and remove biases that might impact your decision-making process.

Once you have metrics established and start measuring various aspects of the API experience and usage, you can see which aspects of your product strategy are being addressed sufficiently by your current product initiatives and where you might have gaps. Establishing an analytics strategy will allow you to stitch together the various sets of metrics and draw the big picture of how your product is meeting your customer and business goals.

It is not sufficient to just have metrics set up. It is also important to understand how to evaluate their quality and how to make sure they are extensive and robust. Each metric on its own is only a measure to evaluate if that measure is at an acceptable level or a concerning level is open to interpretation. In some cases, you have industry standards to match that allow you to set targets, but in many cases, there are no clear benchmarks available, and you might have to do your own research to establish acceptable targets.

In *Part 4*, you will learn the ways in which metrics can be analyzed, interpreted, and evaluated to ensure that you remove blind spots and avoid vanity metrics that may not be a true representation of product health.

Short-term and long-term goal setting

Setting short-term and long-term goals is important for product leadership because it helps to establish a clear direction for a product and the team working on it. Short-term goals provide a sense of immediate progress and accomplishment, while long-term goals provide a sense of purpose and a vision for the future.

In the following chapters, you will learn about why setting an analytics strategy is crucial for the success of your product. You will also learn about various analytical techniques for data mining, text analysis, time series analysis, and so on to help you interpret data and gather insights. You will also learn about goal-setting frameworks such as **specific, measurable, attainable, relevant, and time-bound goals** (**SMART**), (**objectives and key results**) (**OKR**), and north-star metrics to establish short-term and long-term goals that align with your product strategy.

You will also learn about strategic storytelling, which is a way of communicating the goals and vision for a product in a way that is engaging and easily understood by stakeholders. By using storytelling to connect the product's goals to the needs and desires of the target audience, product leaders can build support and buy-in for the product. Additionally, storytelling can be used to help team members understand and align with the product vision, which can help foster a sense of shared purpose and motivation within the team.

Strategic roadmapping

Metrics are important in creating a strategic roadmap because they provide a way to measure and evaluate the performance of a business or product and identify areas for improvement. By using metrics and analytics, organizations can make sure that the goals and objectives in the strategic roadmap are in line with how well the business or product is doing right now and that they can be reached.

Some ways that metrics can be used to create a strategic roadmap include the following:

1. **Identifying areas for improvement**: By tracking key metrics, such as customer satisfaction, revenue, and website traffic, organizations can identify areas where performance is lagging and prioritize these areas for improvement in a strategic roadmap.
2. **Setting goals and objectives**: By analyzing historical performance data, organizations can set realistic and achievable goals and objectives for the strategic roadmap. Metrics and analytics can be used to establish benchmarks and track progress toward these goals and objectives.
3. **Identifying trends and patterns**: By analyzing data over time, organizations can identify trends and patterns in customer behavior and market conditions. This information can be used to anticipate future changes and adapt the strategic roadmap accordingly.
4. **Evaluating the effectiveness of initiatives**: By tracking metrics before and after the implementation of specific initiatives, organizations can evaluate the effectiveness of these initiatives and identify which ones should be continued, scaled, or discontinued in the strategic roadmap.
5. **Improving decision-making**: By providing an objective and data-driven view of performance, metrics and analytics can help organizations make better decisions on where to allocate resources and which initiatives to prioritize in the strategic roadmap.
6. **Stakeholder alignment**: By using metrics, you can align the incentives across various teams within the organization and get them to work on common goals. Data provides an objective set of insights that can drive confidence and agreement within different groups and drive collaboration.

Through the chapters of *Part 4*, you will learn about the interpretation of metrics, goal-setting, removing biases from the interpretation of data, and driving alignment across stakeholders. You will also learn about the techniques of strategic storytelling, using metrics to create a data-driven strategy for your APIs. The following chapters will be covered in this part:

- *Chapter 13, Drawing the Big Picture with Data*
- *Chapter 14, Keeping Metrics Honest*
- *Chapter 15, Counter Metrics to Avoid Blind Spots*
- *Chapter 16, Decision-Making with Data*

13
Drawing the Big Picture with Data

In *Part 3*, we learned about the variety of product metrics you can set across infrastructure, product experience, and business metrics. But each metric only shows you a part of the big picture. To understand the full context, you have to view different metrics in relation to each other. Metrics can be used to figure out how the product is doing, find ways to improve it, and set goals for the different teams.

As you now know, **application programming interfaces** (**APIs**) are built, distributed, and supported by a number of different teams. All these teams and stakeholders have their own set of focused goals they track and optimize. However, as the product manager, you need to understand how these metrics relate to each other so that you can understand how product changes impact the metrics that some of the other teams might be following.

In this chapter, you will learn about the various methodologies used for establishing key metrics across organizations and teams, how to establish a baseline, and find ways of benchmarking it. This chapter will dive into how no metric can be seen on its own and must be evaluated in the context of other metrics. This establishes the concept of correlation in metrics. This chapter will also explain how to create metrics clusters so that there is a set of metrics that are seen in relation to each other rather than all metrics at once.

In this chapter, you will learn the following topics:

- Setting the data strategy
- Methods for analyzing data
- Interpreting data
- Goal setting with data

By the end of this chapter, you will be equipped to set a data strategy for your product, analyze the data you gather, and set metrics that you can track across your team and stakeholders.

Setting the data strategy

You have data if you have a product, but you need more than data to optimize and improve your product. Having a data strategy enables you to turn data into value. Your data strategy comprises the tools, methods, and rules that tell you how to handle, analyze, and use data. A data strategy enables you to make data-driven decisions. It also assists you in keeping your data secure and compliant.

Almost every organization collects data in different ways, and a data strategy helps a company manage and evaluate all of this data. It also puts a company in an excellent position to deal with problems, including the following issues:

- Without capture and analysis of appropriate data, you can't support operational decision-making, and this results in slow and inefficient product operations
- Data privacy, integrity, and quality can limit your capacity to evaluate data
- An inadequate grasp of essential business components (clients, supply chain, competitive landscape, and so on) and the processes that keep them running
- A lack of clarity on present business needs (a problem that descriptive analytics can assist in resolving) and goals (which predictive and prescriptive analytics can help identify)
- Inefficient data flow between business units or duplication by various business divisions can cause misalignment among business units

In short, a company that doesn't have a data strategy isn't likely to work well and make money, let alone grow. Setting up a data strategy can be difficult, and you may need more data points to make a complete strategy most of the time. It's important to set up a framework to simplify the process of making a data strategy and make it iterative. Here is a framework for defining, planning, and shaping your data strategy:

1. *Setting business requirements*: Your data strategy should complement and enhance your entire business strategy. Set clear goals and measurable objectives for your data strategy that help your business plan. Your data plan, for example, could maintain data storage costs below a given threshold. For your data strategy to help you reach your goal of controlling data storage costs, it would need to define storage solutions or services that meet your cost criteria and best practices to help users optimize storage costs. It should also define indicators, such as average cost per gigabyte of storage, to assist you in tracking your progress toward this target.
2. *Establish both long-term and short-term objectives*: You might set a short-term goal of reviewing data quality once a month. Still, your long-term goal might be continuous data quality, which means finding and fixing data quality problems all the time instead of relying on checks.
3. *Sourcing the data*: Take the time to identify your existing data sources, noting where the data is located and how it is obtained. Identify data gaps and barriers to acquiring the data you require. Consider using third-party data, such as demographic, market, competitor, and geospatial information, to understand your internal data comprehensively.

4. *Designing data infrastructure*: Focus on the business priorities for your data projects rather than the hype and latest technologies. Creating a data architecture that is strong, easy to use, and scalable takes time and can be done in many ways. And, as with most designs, the more your requirements and future demands are taken into account, the better the solution will assist the business and adapt to your needs. Product managers in large organizations often rely on technological partners such as infrastructure teams, business intelligence teams, and so on to make data infrastructure decisions.
5. *Turning data into insights*: A data visualization tool is essential for obtaining vital business insights. The data analysis and visualization tool you choose should make it easy to understand the data it gives you. However, just like when assessing your technology infrastructure requirements, there are extra considerations when selecting your tool, such as its capacity to accommodate existing data. In most organizations, the decision for data tools is made by the database architecture and analytics teams at the company level, so you may not have to decide the tooling needs for each new product.
6. *Visualizing the data*: Looker and Tableau are the most popular tools for visualizing data. They let people from all over an organization access, analyze, and build dashboards that can be shared and used repeatedly.
7. *Identifying people and processes*: People in your business, as well as methods for collecting, sharing, and regulating data, are just as vital as technology. Assess skill sets to identify strengths and areas where users require assistance. Do the engineering, operations, and sales teams have the tools and support from leadership they need to use data the way you want them to? Is training required? Are you prepared to meet the demands of the business? Business processes may need to be redesigned to include data analysis and eliminate mistakes that make it hard to use data.
8. *Establishing data governance*: A data governance program gives users access to verified, high-quality information and explains the available data and how to use it. A data governance program must be realistic to be successful—it must meet your goals, urgency, maturity, and capabilities and be structured to scale as needed. The data governance processes will make your data and analytics more accessible, trustworthy, and actionable.
9. *Roadmapping*: Improving data capture is an ongoing aspect of building products. You may never have all the data you need, and you will continuously have to instrument data points to be captured to support decision-making about the product. Your product roadmap must accurately represent ongoing work to capture data and enable analytics. This requires you to plan data just the way you plan product features. This comprehensive plan identifies and prioritizes all the activities that must occur to get you from where you are to where you want to go across data engineering and analytics, which might often span multiple teams.

Once you have set up your data strategy, you need to keep making changes to it to fill in any gaps you find in data collection and analysis. Having data is only half the battle; the next step is to start analyzing it and turning it into insights. In the next section, you will learn about methods of data analysis that will help you explore and interpret data.

Methods for analyzing data

Different types of data need to be analyzed differently. For example, usage metrics need to be analyzed in relation to a time period where you can track how customers use your product over time. But when you start to dive into customer behavior, you should segment the user base and understand how the different clusters of customers behave in contrast with each other.

In this section, you'll learn the 10 most important methods for analyzing data and how to use them to set up API product analytics.

Cluster analysis

Cluster analysis is a way to use statistics to find groups of similar observations in a set of data. It is a way to break up a large set of different data into smaller, more similar groups based on patterns and relationships in the data. The following screenshot shows the plot of customers across the number of developers on their team on the *x* axis and the time to the first Hello World metric on the *y* axis. In this example, you can see that there are clusters forming in the plot, showing that **small and medium-sized businesses** (**SMBs**) tend to be in a similar range for these two variables.

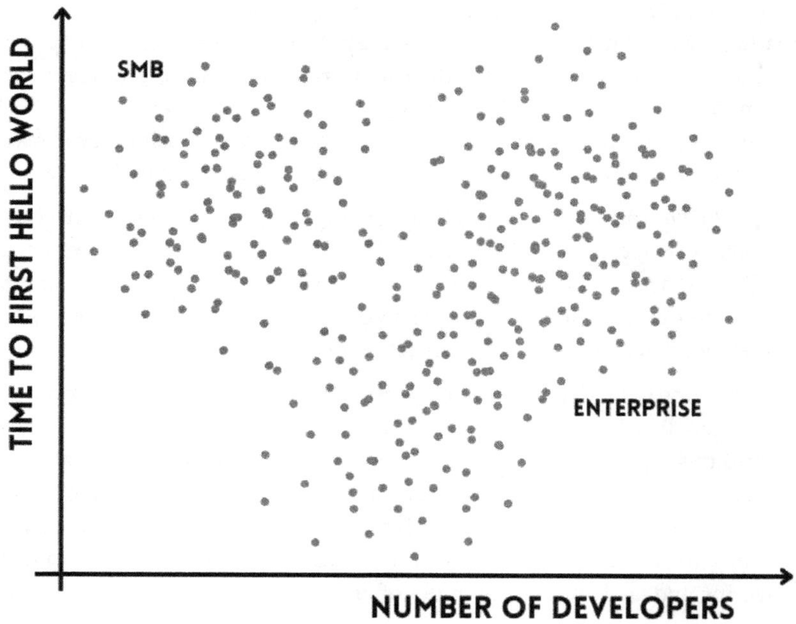

Figure 13.1 – Cluster analysis showing clusters of enterprise and SMB customers

Cluster analysis can be used to find groups of customers with similar traits or patterns of behavior when API data is being looked at to understand customer segmentation. For example, the API data might include information about the types of products or services customers use, their geographic location, demographic characteristics, or other factors relevant to understanding customer behavior. By applying cluster analysis to this data, it is possible to identify customers with similar characteristics or behavior patterns and understand how these groups differ.

Cluster analysis can be used on API data to understand customer segmentation. One way to do this is to use a clustering algorithm to find groups of customers with similar traits or patterns of behavior. Many different clustering algorithms can be used for this purpose, each with its own strengths and limitations.

There are many factors to consider when deciding which clustering algorithm to use to understand customer behavior. Some of the factors that might influence your choice are listed here:

- *Size of the data*: If you have a huge dataset, you may want to use a clustering algorithm that is computationally efficient.
- *The nature of the data*: Some clustering algorithms are better suited for certain data types than others. For example, density-based clustering algorithms are good for finding clusters in datasets with a non-uniform distribution.
- *The number of clusters you expect to find*: Some clustering algorithms, such as k-means, require you to specify the number of clusters you expect to find in advance. Others, such as hierarchical clustering, do not need you to specify the number of clusters.
- *The interpretability of the results*: Some clustering algorithms, such as K-means, produce results that are relatively easy to interpret. Others, such as spectral clustering, may produce results that are more difficult to interpret.

In the end, which clustering algorithm you use will depend on the goals and limits of your analysis and how familiar you are with the different ones. Once the clusters have been found, their characteristics can be compared to determine what makes them different. This information can then be used to develop targeted marketing or product development plans.

Cohort analysis

Cohort analysis is a statistical method used to find patterns and trends among people who have something in common, such as customers of the same business or users of the same product. Cohort analysis can be used to figure out how different groups of users interact with different pages of the API documentation. This is done in the context of analyzing API references.

To do cohort analysis on API reference pages, you first need to find what all the users in the group you want to study have in common. For example, you could say that your cohort is all the users who went to a particular API documentation page during a specific time period. You would then collect data on how these users interacted with the API documentation, such as which pages they accessed, how long they spent on each page, and whether they completed any actions (such as making an API call or signing up for an account).

Once you have this information, you can use different statistical methods to find patterns and trends in the cohort. For example, you might look for differences in how other groups of users interact with the API documentation, or you might look for changes in user behavior over time. By understanding how different groups of users interact with the API documentation, you can identify areas for improvement and make changes to meet the needs of your users better.

The cohort analysis method can be used with tools for web analytics such as Google Analytics and Heap Analytics. The screenshot in *Figure 13.2* is an example of visualizing a cohort using Google Analytics. It shows user segmentation by date cohorts (**Acquisition Date**) and then evaluated week by week to extract performance insights.

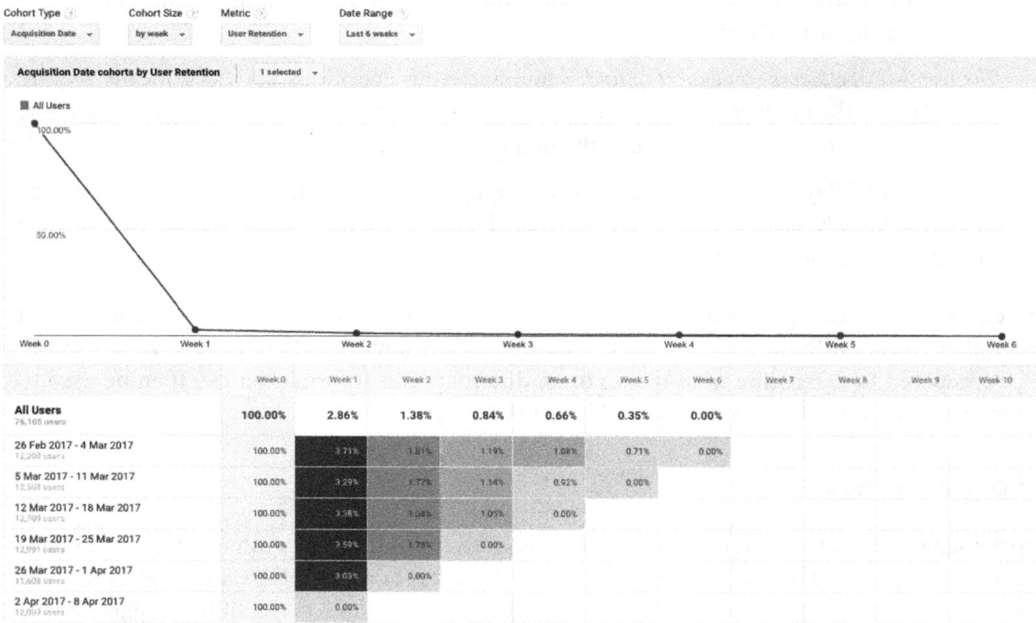

Figure 13.2 – Google Analytics cohort analysis report

Cohort analysis can be valuable for API documentation pages in several ways:

- *Identifying user needs*: By analyzing how different groups of users interact with the API documentation, you can identify their needs and preferences. This can help you better tailor the documentation to meet your users' needs.
- *Improving user experience*: By identifying patterns and trends among different groups of users, you can identify areas of the API documentation that are causing confusion or frustration for users. This can help you improve the documentation to make it more user-friendly.
- *Identifying areas for growth*: By analyzing the behavior of different groups of users, you can identify areas of the API documentation that are particularly popular or successful. You can then focus on promoting and expanding these areas to drive further growth.
- *Measuring the effectiveness of changes*: By performing cohort analysis before and after making changes to the API documentation, you can measure the effectiveness of those changes and determine whether they have the desired impact on user behavior.

Overall, cohort analysis can help you understand how different groups of users interact with the API documentation and identify opportunities for improvement, which can ultimately lead to a better user experience and more successful adoption of your API.

Regression analysis

Regression analysis is a statistical method used to find out how different variables are related to each other. It can be used to predict the value of one variable (the dependent variable) based on the values of one or more other variables (the independent variables).

When evaluating API products, regression analysis can determine how different factors, such as pricing, performance, and usability, affect success or adoption measures, such as revenue, number of users, or usage. By looking at these connections, you can determine which factors are most important for success and decide how to improve your API product intelligently.

To use regression analysis to evaluate an API product, you first need to collect data on the variables you are interested in. This might include data on the API's pricing, performance, and usability, as well as data on metrics such as revenue, number of users, and usage. Then, you would use statistical software to fit a regression model to the data and examine how the variables are related.

For example, you might use regression analysis to understand how changes in the pricing of your API product are related to changes in the number of users or revenue. This can help you figure out how to price your API product in the best way and make intelligent choices about how to price it in the future.

Predictive analysis

Predictive analysis is a statistical technique used to predict future events or outcomes based on historical data. It can be used to identify patterns and trends in data and use those patterns to make informed guesses about what is likely to happen in the future.

In evaluating API products, predictive analysis can be used to forecast metrics such as revenue, the number of users, or usage. This can help you identify trends and make informed decisions about improving your API product.

To use predictive analysis to evaluate an API product, you would first need to collect data on the variables of interest. This might include data on the API's pricing, performance, and usability, as well as data on metrics such as revenue, number of users, and usage. Using techniques such as linear regression or time series analysis, you would then use statistical software to build a predictive model based on this data.

For example, you could use predictive analysis to estimate how many people will use your API product based on past pricing, performance, and ease of use data. This can help you find patterns in how users use your API product and make smart decisions about how to improve it to drive more growth.

Data mining

Data mining is the process of discovering patterns and knowledge from large sets of data through the use of algorithms and statistical models. It is often used to identify patterns that can be used to make decisions or predictions.

API usage can be evaluated using data mining analysis by analyzing the data generated by API usage. This can include things such as the number of API requests made, the amount of data transferred, the types of API calls being made, and the success or failure rate of API calls. By analyzing this data, you can identify trends and patterns in API usage, such as which API calls are most popular, experiencing the most errors, and using the most resources. This information can be used to optimize API usage and improve the overall performance of the API.

Imagine that you are the product manager of an API that provides weather data to app developers. You want to use data mining analysis to understand how the API is being used and to identify any trends or patterns that can help you optimize the API.

To begin the analysis, you would gather data on the API usage, including the number of API requests made, the amount of data transferred, the types of API calls being made, and the success or failure rate of API calls. This data might be stored in a database or log files.

Next, you would use data analysis and visualization to look at the data and find trends and patterns. For example, you could use statistical analysis to find out which API calls are the most popular, have the most errors, and use the most resources. You might also use **machine learning** (**ML**) techniques to predict which API calls are likely to be made in the future based on the patterns that you have identified in the data.

With this information, you can make informed decisions about how to optimize the API. For example, you may focus on improving the performance of the most popular API calls, or you may decide to add additional resources to handle the API calls that are experiencing the most errors. In general, data mining analysis can help you figure out how your API is being used and find ways to make it better.

Text analysis

Text analysis is a way to get information from text data by using techniques such as **natural language processing** (**NLP**) and ML. It is often used to find patterns and trends in text data and to classify, cluster, and summarize text data.

Text analysis can be used in a number of ways to look at data about how APIs are used. For example, you could use text analysis to do the following:

- Analyze the text data generated by API usages, such as error messages, log files, and user feedback. In turn, this could help you identify trends and patterns in API usage, such as which API calls are most popular, are experiencing the most errors, and are using the most resources.
- Classify API usage data into different categories based on the content of the text data. For example, you might classify API calls as successful or unsuccessful based on the text of the error messages that are generated.
- Cluster API usage data into groups based on the similarity of the text data. This could help you identify groups of API calls that are similar in some way, such as API calls that are experiencing similar errors or being used by similar types of users.
- Summarize API usage data by extracting key points or trends from the text data. This could help you quickly understand the most critical aspects of the API usage data, such as the most common types of API calls or the most common reasons for API failures.

Overall, text analysis can be a powerful tool for understanding and evaluating API usage data because it lets you pull insights and trends from large amounts of text data.

Time series analysis

Time series analysis is a method of analyzing data collected over time. It is used to understand trends and patterns in data that change over time, such as stock prices, weather data, and population demographics.

By looking at how the data changes over time, time series analysis can be used to evaluate data about how APIs are used. For example, you could use time series analysis to do the following:

- Identify trends and patterns in API usage over time. For example, look at the number of API requests made per day, week, or month, and see whether there is a regular pattern to the user or whether there are sudden spikes or dips.

- Forecast future API usage based on past trends. By analyzing the patterns in API usage over time, you can build a model that predicts how much the API will be used in the future. This can be helpful for resource planning and capacity management.
- Identify unusual or unexpected behavior in API usage. By comparing API usage data to past trends, you can identify anomalies that may indicate a problem with the API or with the way it is being used.

To perform time series analysis on API usage data, you would typically need to gather data on API usage over time and then use statistical techniques such as time series modeling and forecasting to analyze the data. With the insights gained from the analysis, you can make informed decisions about optimizing the API and managing its usage.

Decision trees

Decision tree analysis is a method of analyzing data that involves creating a tree-like model of decisions and their possible consequences. It is often used in ML and data mining to use data to predict what will happen.

Decision tree analysis can be useful for API usage data in a number of ways. For example, you could use decision tree analysis to do the following:

- Predict the likelihood of API failures based on past data. By analyzing past API usage data and the outcomes of those API calls (for example, success or failure), you can build a decision tree model that predicts the likelihood of API failures based on specific characteristics of the API calls (such as the type of API call and the data being passed to the API).
- Identify the factors that contribute to API failures. By building a decision tree model based on past API usage data, you can identify the factors most strongly associated with API failures. This can help you understand why the API is failing and what you can do to prevent future failures.
- Optimize API usage by prioritizing API calls. By analyzing the characteristics of API calls that are most likely to succeed or fail, you can prioritize API calls based on their likelihood of success. This can help you optimize the API by focusing on the API calls that are most likely to be successful and minimizing the number of API calls that are likely to fail.

To perform decision tree analysis on API usage data, you would typically need to gather data on the API usage and the outcomes of those API calls and then use ML techniques to build and evaluate a decision tree model based on that data.

Conjoint analysis

Conjoint analysis is a method of analyzing data to understand how people value different product or service features. It is often used in market research to help businesses understand what features are most important to consumers and how much they are willing to pay for them.

Conjoint analysis can be used for analyzing API usage data in several ways. For example, you could use conjoint analysis to do the following:

- Determine the optimal pricing for the API. By understanding how users value different API features, you can determine how much users are willing to pay for those features and set the pricing for the API accordingly.
- Identify opportunities for new features or improvements to the API. By understanding which API features are most valuable to users, you can identify opportunities for new features or enhancements that will be most appealing to them.

To perform conjoint analysis on API usage data, you would typically need to gather data on the API usage and user feedback and then use statistical techniques such as multivariate analysis to analyze the data and understand how users value different API features.

Imagine that you are the product manager for an API that provides financial data to app developers. You want to use conjoint analysis to understand how users value different API features and determine the optimal pricing for the API.

To begin the analysis, you would gather data on the API usage and user feedback, such as the number of API requests made, the types of API calls being made, and any comments or ratings provided by users. You would then identify the API features you want to analyze, such as the breadth of financial data available, the frequency of updates to the data, and the level of support the API team provides.

Next, you would use statistical techniques such as multivariate analysis to analyze the data and understand how users value each of the features of the API. For example, you might find that users are willing to pay more for a wider range of financial data but are less concerned about the frequency of updates to the data.

With this information, you can determine the optimal pricing for the API by setting prices that reflect the value that users place on different features of the API. You can also use the insights gained from the conjoint analysis to identify opportunities for new features or improvements to the API that will be most appealing to users.

Factor analysis

Factor analysis is a statistical method used to identify the underlying factors or dimensions that explain the variance in a dataset. It is often used in psychology, sociology, and marketing to understand the relationships between variables and to identify patterns in data.

Factor analysis can be useful for API usage data in a number of ways. For example, you could use factor analysis to help you to do the following:

- Identify the underlying factors that influence API usage. By analyzing data on API usage, you can use factor analysis to identify the factors most strongly associated with API usage. This could include the type of API calls being made, the data being passed to the API, and the characteristics of the users making the API calls.
- Understand the relationships between different variables in the API usage data. By identifying the underlying factors that influence API usage, you can understand how different variables in the data are related to each other and how they influence API usage.
- Simplify and summarize the API usage data. By identifying the underlying factors that explain the variance in the data, you can reduce the complexity of the data and focus on the most critical factors that influence API usage.

To perform factor analysis on API usage data, you would typically need to gather data on the API usage and the variables that you want to analyze and then use statistical techniques such as principal component analysis or common factor analysis to identify the underlying factors in the data.

Imagine that you are the product manager for an API that provides movie recommendation data to app developers. You want to use factor analysis to understand the underlying factors that influence API usage and identify data patterns.

To begin the analysis, you gather data on the API usage over a period of time, including the number of API requests made, the types of API calls being made (for example, recommendations for a specific genre of movies, recommendations based on a specific actor or director), and the characteristics of the users making the API calls (for example, age, gender, location).

You then use factor analysis to identify the underlying factors influencing API usage. You find that three main factors influence API usage:

- The type of API calls being made
- The age of the users making the API calls
- The location of the users

With this information, you can understand the relationships between different variables in the API usage data and how they influence API usage. For example, users in specific locations are more likely to make API calls for recommendations for specific genres of movies, or users of certain ages are more likely to make API calls for recommendations based on specific actors or directors.

You can also use the results of the factor analysis to simplify and summarize the data, focusing on the most important factors that influence API usage. This can help you understand how the API is used and identify improvement or optimization opportunities.

Interpreting data

Data interpretation is the process of understanding and making sense of data. It involves examining and drawing conclusions based on the data. The data analysis and visualization process involves gathering data, creating visualizations, performing analysis, and drawing actionable insights, as shown in the following diagram:

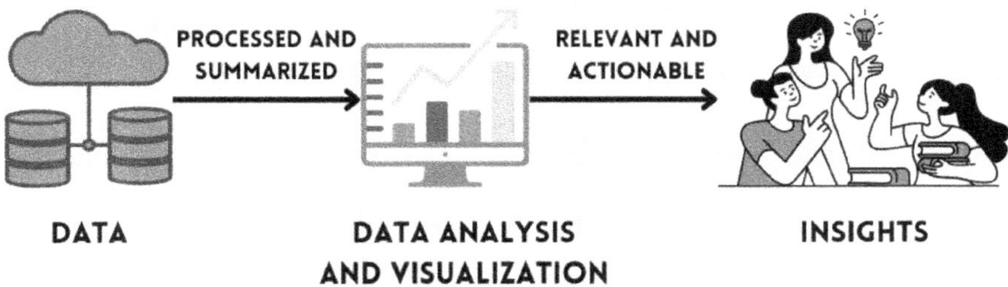

Figure 13.3 – The process of interpreting data to form insights

1. *Summarize and process the data*: The first step in the data interpretation process is to summarize and process the data in a way that makes it easier to understand. This might involve organizing the data into a tabular format, calculating summary statistics, or creating pivot tables.
2. *Create visualizations*: Once the data has been summarized and processed, it is often helpful to create visualizations to help understand the data and identify patterns and trends. Visualizations can include charts, graphs, maps, or other types of diagrams.
3. *Perform visual analysis*: After creating visualizations, the next step is to perform a visual analysis to identify patterns, trends, and relationships in the data. This might involve comparing different variables, identifying outliers, or looking for correlations between variables.
4. *Draw insights*: The final step in the data interpretation process is to draw insights from the data. This might involve making conclusions about the data, identifying trends or patterns, or formulating hypotheses about the relationships between variables.

Overall, the process of data interpretation involves summarizing and processing data, creating visualizations, performing visual analysis, and drawing insights from the data. This process helps make sense of large and complex data sets and informs decision-making.

Product managers need to know how to interpret data because it is a key source of information for making smart product development and strategy decisions. By analyzing data, you can gain insights into how your products are being used, what features are most valuable to users, and how to optimize the product to meet the market's needs.

Product managers can use both qualitative and quantitative data interpretation techniques to make product decisions. In **qualitative data interpretation**, you look at data that isn't a number, such as customer comments, user interviews, and transcripts of focus group discussions. This kind of data can tell you a lot about how users think, feel, and act, which can help you decide which features to put first and how to design the user experience.

Quantitative data interpretation involves analyzing numerical data, such as usage statistics, market research, and financial data. This type of data can provide objective, measurable insights into how a product is being used, what features are most popular, and how the product is performing financially. **Statistical methods** such as **regression analysis** and **hypothesis testing** can be used to look for trends and patterns in quantitative data.

Product managers can use several general rules to ensure they make well-informed, fair decisions when interpreting data. These guidelines include the following:

- *Avoiding confirmation bias*: Confirmation bias is the tendency to interpret data to confirm one's preexisting beliefs or hypotheses. To avoid this bias, it is important to objectively approach data and be open to the possibility that the data may contradict one's initial assumptions.
- *Understanding causation versus correlation*: It is important to understand the difference between causation and correlation when interpreting data. Correlation means that two variables are related in some way, but it does not necessarily mean that one variable causes the other. Causation means that one variable causes the other. It is important to consider other potential explanations for the relationship between the variables and to use experimental methods to test the relationship when determining causation.

- *Understanding statistical significance*: Statistical significance refers to the likelihood that a relationship between two variables is not due to chance. A relationship is considered statistically significant if the probability of observing the relationship by chance is very low (usually less than 5%). It is essential to consider statistical significance when interpreting data, as it helps to distinguish between meaningful relationships and random fluctuations in the data.

By following these rules, product managers can make sure they are objectively and correctly interpreting data and making decisions about their products and strategies based on facts.

In the following section, you will learn about the SWOT framework, which is a quick way to evaluate and prioritize the current state of the product.

SWOT analysis

Strengths, weaknesses, opportunities, and threats (**SWOT**) analysis has become popular because it is quick, easy, and gives a quick overview of the situation of the product. A SWOT analysis can help find problems and determine if the goal can be reached in the current operating environment. The results of a SWOT analysis can be shown as a simple list or as a series of columns. The most common way to show this is with a matrix like this:

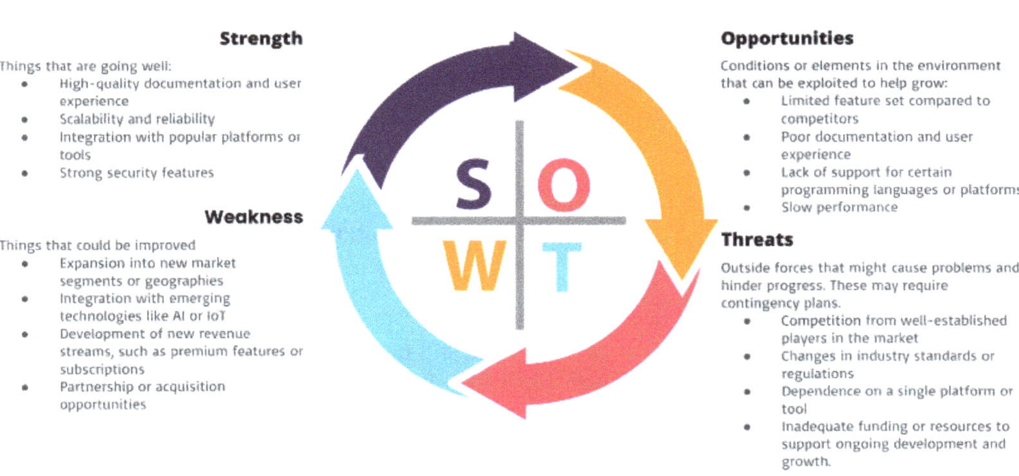

Strength
Things that are going well:
High-quality documentation and user experience
Scalability and reliability
Integration with popular platforms or tools
Strong security features
Weakness
Things that could be improved
Expansion into new market segments or geographies
Integration with emerging technologies like AI or IoT
Development of new revenue streams, such as premium features or subscriptions
Partnership or acquisition opportunities
S
O
W
T
Opportunities
Conditions or elements in the environment that can be exploited to help grow:
Limited feature set compared to competitors
Poor documentation and user experience
Lack of support for certain programming languages or platforms
Slow performance
Threats
Outside forces that might cause problems and hinder progress. These may require contingency plans.
Competition from well-established players in the market
Changes in industry standards or regulations
Dependence on a single platform or tool
Inadequate funding or resources to support ongoing development and growth.

Figure 13.4 – SWOT analysis matrix

This is how the SWOT method works:

1. Make your goal very clear.
2. Find your strengths or the things you do well that could help you reach the goal.
3. Find your weaknesses—areas that need work and could hold you back.
4. Find opportunities or places that are good for growth or will help you in the future.
5. Find the things that could hurt your efforts, such as competitors or situations.
6. Figure out how the things you've named fit together.
7. Cut down on the things you want to talk about and put them in order of importance.

The parts of a SWOT analysis can be very different, but the analysis must be realistic and thorough. SWOT is a tool for making plans. It's about making plans for the future, so pay attention to things that could affect whether you reach the goal.

When figuring out the pros and cons of your product, you should look at how it compares to competitors and industry standards. You can do this by using benchmarking and baseline techniques, which you will learn about in the next section.

Benchmarking and baselining

A **baseline** for your performance metric is the average level of performance that your product is currently at. You will compare future performance levels to this average level to see if performance changes.

Most of the time, benchmarks for measuring performance come from outside your organization. They are levels of performance that another organization has reached for the same performance measure that you are interested in. For example, benchmarks from organizations that show the best practices in the world or industry can be used as goals for measuring performance. Because other people have done that well, it's easier to think you can.

Even if you have a standard, your business processes can still meet it. To improve performance, we need to ensure that all of our processes can meet any standard. Standards are like goals, but they apply to the *here and now*. You've committed to your customers and want to see how well you're keeping that promise.

Establish a baseline for the metrics you establish to measure your product so that you can evaluate them as they change over time. Although it can be challenging to find external benchmarks for all metrics, you can occasionally reverse-engineer these metrics for competitors as part of a competitive study.

Developing a deep understanding of your metrics in the context of baseline and benchmarks enables you to make effective decisions with data that you will learn about in the next section.

Making decisions without data

In an ideal world, all the data would be captured accurately and would be readily accessible for analysis. However, because of the tremendous amounts of data being generated, it is impossible to structure or organize data. Additionally, data is often scattered over several different systems in large organizations, and gathering data for analysis is a challenge. Scenariosinvolving facing the challenge of making decisions without data include the following:

- *Time constraints*: In some cases, there may need to be more time to gather and analyze data before a decision needs to be made. In these cases, the ability to make decisions with insufficient data can be crucial for making timely decisions.
- *Complexity*: Data-driven decision-making may only sometimes be possible or appropriate in situations that are too complex to analyze with existing data or tools. In these cases, the ability to make decisions with insufficient data can be necessary for finding solutions to complex problems.
- *Changing circumstances*: In rapidly changing environments, data may only sometimes be available or quickly become outdated. In these cases, making decisions with insufficient data can help organizations be more flexible and responsive to changing circumstances.
- *Expertise*: In some cases, the expertise and experience of individuals can be more valuable than data in making decisions. For example, a highly experienced surgeon can make better decisions about a patient's treatment than a computer program, even if the computer program has access to more data.

Overall, being able to make decisions when you don't have enough information can be helpful in a number of situations. However, it's important to be aware of this method's limitations and to use it in conjunction with data-driven methods and directional insights whenever possible.

Directional insights are insights or indications that provide a sense of the direction or trend of something without providing factual data or measurements. Directional insights can be helpful in making decisions when concrete data is unavailable, as they can help provide a general sense of the direction the data is likely to take.

There are several ways to use directional insights to make decisions when factual data is unavailable, including the following:

- *Expert opinion*: Seeking experts' opinions in a particular field can provide valuable directional insights, as experts can draw on their knowledge and experience to provide a sense of the direction the data is likely to take
- *Market research*: Conducting market research, such as focus groups or surveys, can provide directional insights into the preferences and behavior of consumers
- *Benchmarking*: Comparing the performance or characteristics of a product or service to those of similar products or services can provide directional insights into how the product or service is likely to perform

- *Trend analysis*: Analyzing trends in data over time can provide directional insights into how the data is likely to change in the future

By using these approaches, product managers can gain valuable directional insights that can inform their decision-making, even when concrete data is not available. However, it is important to recognize that directional insights are not concrete data and may not always be accurate. As such, it is important to consider the limitations of this approach and use it in conjunction with data-driven approaches whenever possible.

In this section, you learned about a number of approaches that can be used to make decisions without data, such as intuition, expert opinion, market research, benchmarking, and trend analysis. Another noteworthy method for decision-making is the Delphi method, which you will learn about in the next section.

The Delphi method

The Delphi method is a structured process for gathering and aggregating the opinions of a group of experts. It involves conducting a series of anonymous rounds of voting and feedback, in which the experts are asked to provide their opinions on a particular topic. The results of the Delphi method can be used to make decisions when data is unavailable or when the subject matter is complex and requires the input of multiple experts.

Imagine that a company is thinking about releasing a new product, but they need more information before it can decide whether or not to do so. The company decides to use the Delphi method to gather the opinions of a group of experts to make a decision.

The Delphi method involves conducting a series of anonymous rounds of voting and feedback in which the experts are asked to provide their opinions on the new product. In the first round, the experts are asked to provide their initial opinions on the product, along with any relevant data or information that they have. In subsequent rounds, the experts are asked to review the opinions and data provided by the other experts and to revise their own opinions as necessary.

After several rounds of voting and getting feedback, the company can get a wide range of opinions and ideas from experts. They can then use these to decide whether or not to move forward with the new product. While the data gathered through the Delphi method may not be concrete, it can provide valuable directional insights that inform decision-making.

Overall, the Delphi method is a structured way to get the opinions of a group of experts and put them all together. It can be helpful when there needs to be more information or when the situation is complicated or specialized.

In API products, these methodologies can be helpful in making decisions when data is unavailable or when the situation is too complex to analyze. But it's essential to keep in mind that these methods have their limits and to combine them with data-driven methods whenever possible. While analysis and interpreting of data allow you to gain insights into your customers as well as product performance, using these insights, you can begin to set goals for your product and your team using goal-setting frameworks discussed in the next section.

Goal setting with data

If you and your team are being pulled in so many different directions that you don't know your goals or how to reach them, it's time to take a step back. Review your product strategy and ensure that your goals are realistic, measurable, and in line with the product vision. To make sure you're setting the right goals, you can use one of the following frameworks for goal-setting.

SMART framework

The **setting specific, measurable, attainable, relevant, and time-bound** (**SMART**) goal-setting framework is defined as follows:

- **Specific**: Specify clearly what you intend to do, who is in charge of it, and what activities must be completed. Please be as specific as possible. Specificity of goals is critical in this process because having ambiguous goals can be not only difficult to achieve but also hard to measure and track.
- **Measurable**: Each goal should have a measurable outcome and a clear definition of success. This will allow you to assess your accomplishments and track your progress. You won't know whether or not you succeeded if you can't measure it.
- **Achievable**: Outline the measures your team must take to achieve the goals and ensure they are realistic and reasonable. There will surely be obstacles along the path, so ensure they are identified ahead of time.
- **Relevant**: Keep your product vision in mind and set targets that will be worth the effort. Knowing why each of the goals you establish is significant is specific.
- **Time-bound**: Set a deadline for each goal. Set start and end dates so everyone on your team is on the same page. Otherwise, you risk discouraging your crew. Include check-in dates to revisit goals as you go and determine whether you should iterate.

The methodology for establishing SMART goals can be summed up in trying to write down your goals in the following format:

Our objective is to [quantifiable goal] **by** [deadline].
[Team members] **will achieve this aim by doing** [actions to attain the goal].
Achieving this aim will benefit us [outcome or advantage].

Figure 13.5 – SMART framework format of writing goals

The OKR framework

The **objectives and key results** (**OKR**) framework is a goal-setting method that is used by organizations to set and achieve measurable objectives. It is a flexible and agile method that involves setting ambitious goals (objectives) and measurable targets (key results) to track progress toward those goals.

The OKR framework is meant to help people and teams align their goals with the organization's overall goals. It is based on the idea that by setting clear and measurable goals, organizations can increase focus, alignment, and engagement and drive continuous improvement.

In the OKR framework, objectives are high-level goals that are intended to inspire and challenge individuals and teams. Key results are specific and measurable targets used to track progress toward the objectives. Key results should be quantifiable and objective and should be used to measure the success or failure of the objectives.

OKRs are normally written with an objective at the top and three to five supporting key results beneath. OKRs can also be written as statements, as in the following example:

I will [objective] **as measured by** [Key Results].

Figure 13.6 – The OKR framework format of writing goals

An objective is what has to be accomplished—nothing more and nothing less. Objectives are defined as meaningful, concrete, action-oriented, and (hopefully) motivating. When designed and used correctly, they protect against fuzzy thinking and poor execution.

Key results (**KRs**) are used to set a baseline for measuring progress toward the goal. Effective KRs are specific, time-bound, and aggressive while remaining realistic. Most importantly, they are measurable and verifiable. There is no gray area or space for doubt; you either satisfy or do not fulfill the requirements of a key result. At the end of the given term, which is usually a quarter, you can do a periodic check to see whether the critical goals have been met or not.

Even though an objective can be stretched out for a year or more, key results should change as the work goes on. The objective is fulfilled once all of them are completed.

Key performance indicators (KPIs)

Key performance indicators (**KPIs**) are the quantitative criteria used to determine long-term strategic, financial, and operational business performance. The KPI framework is often built around a set of objectives, targets, and industry benchmarks.

KPIs are classified as follows:

- **Customer-centric**: This includes customer **lifetime value** (**LTV**), **customer acquisition cost** (**CAC**), **customer satisfaction score** (**CSAT**), **customer satisfaction score** (**CES**), and **net promoter score** (**NPS**), as well as customer experience and retention
- **Process-centric**: This includes operational performance indicators, business intelligence, and product analytics
- **Financially-focused**: This includes indicators such as **monthly recurring revenue** (**MRR**), **annual recurring revenue** (**ARR**), gross profit margin, bottom line, and other financial metrics

KPIs are more commonly used than OKRs by SaaS businesses, startups, and small organizations.

North Star metrics

The North Star metric is a framework that helps companies identify the most important metric that they should track to measure the success of their business. It is a single metric that represents the core value that the company delivers to its customers and that aligns the entire company around a common goal. This framework is commonly used in startups and digital product companies but can be applied to any business.

The North Star metric can be applied to an API product by identifying the core value that the API delivers to its customers and then selecting a metric that best represents that value. For example, if the core value of the API is to provide accurate and reliable data to its customers, a possible North Star metric could be the percentage of successful API calls. This metric would align the entire company around the goal of ensuring that the API is delivering accurate and reliable data to its customers. Another example could be if the core value of the API is to allow for easy and efficient data integration; a possible North Star metric could be the number of active integrations. This metric would align the entire company around the goal of ensuring that the API is easy to integrate with other systems and that more and more customers are using it.

Your North Star metric will differ depending on your team structure, product life cycle stage, and other factors, but it should always be unique to your organization, embodying your product vision and mission.

It should be noted that the North Star Metric is not something you choose only once. The most important metric for your business will alter and change as your firm expands and your product gets closer to the product vision. Overall, remember that your North Star metric must be a leading sign of future performance. It should also be well defined, easily understandable, and actionable by everyone on your team.

The HEART framework

After finding *new problems and opportunities for large-scale assessment of user experience*, Google made the HEART framework. Google says that HEART is a framework for user-centered metrics and a way to connect product goals with metrics. This can assist your product team in making data-driven and user-centered decisions.

HEART is an acronym that stands for:

- Happiness
- Engagement
- Adoption
- Retention
- Task success

The HEART framework could be a great way to look at goals and measure user experience across the board, from how people use a single feature to how happy they are with your product as a whole. Google also encourages teams to perform additional studies using enormous amounts of behavioral data.

We recommend that, in addition to product statistics, you use microsurveys to get feedback from users on a regular basis and make your product more customer-focused as you go.

Summary

In this chapter, you learned how to set up a data strategy that lets you collect the right data points and make an iterative plan for building a lot of analytics for your product. You also learned ways to look at data, such as cluster analysis, data mining, predictive analysis, and so on. This gave you the tools you need to start exploring data.

Setting a data strategy and using various methods for analyzing and interpreting data can be extremely helpful in effective goal setting with data. A data strategy is a plan that outlines how an organization will collect, analyze, and use data to support decision-making and drive business objectives. By setting a data strategy, organizations can ensure that they are collecting and analyzing the right data to inform their goals and objectives.

There are many different methods for analyzing and interpreting data, such as statistical analysis, ML, and visual analysis. By using these methods, organizations can gain insights and understand trends and patterns in their data and use this information to inform their goals and objectives.

Overall, setting a data strategy and using various methods for analyzing and interpreting data can help organizations to set clear, measurable, and data-driven goals and to track progress toward those goals. This can increase focus, alignment, and engagement and drive continuous improvement toward the organization's objectives.

Setting up a data strategy along with your product strategy is the first step toward becoming data-driven. As you approach new data and start to analyze it, you can baseline and benchmark your findings to make sure you remove bias from your interpretation as you identify strengths, weaknesses, opportunities and threats for your product strategy. You can use all these insights to set goals using one of the five goal-setting frameworks you learned in this chapter to get your team and stakeholders aligned and working together to accomplish these goals.

14
Keeping Metrics Honest

In *Chapter 13*, *Drawing the Big Picture with Data*, you learned about analysis methods and goal setting with data. It is impossible to implement and track all possible metrics at all times. Setting a high-level strategy and goals allows you to identify and prioritize the right metrics for initiatives in the short term.

When we talk about data, we often only think about quantitative data and lose sight of qualitative data. Qualitative and quantitative data should be combined to form hypotheses and drive insights that may not be easily available without combining these two. Creating clusters of metrics and constantly validating hypotheses based on findings from one perspective with another set of metrics or qualitative insights allows you to remove biases from your metrics. The topics covered in this chapter include the following:

- Mixing qualitative and quantitative feedback
- Validating your insights
- Defining the right product metrics
- Framework for storytelling with data

By the end of this chapter, you will have learned about effective storytelling with data by combining qualitative and quantitative data and creating clusters of metrics to create a good coverage of metrics.

Mixing qualitative and quantitative feedback

Quantitative metrics provide objective and numerical data, which can help eliminate bias and subjectivity. By analyzing these metrics, you can make informed decisions about the performance and efficiency of your API without being swayed by personal opinions or preferences.

Qualitative metrics, on the other hand, provide a more subjective evaluation of the API and can help identify areas that may not be captured by quantitative metrics. For example, customer feedback may highlight issues or concerns that are not reflected in numerical data. By gathering this type of data, you can get a better understanding of how an API is being used and perceived by your customers and address any potential blind spots or biases. Qualitative and quantitative data work well together because each kind fills in what the other lacks.

You might see patterns in the numbers you already track. For example, most of your customers don't use a particular feature. Then, you could add qualitative research, such as one-on-one interviews, to find out why customers aren't using that feature. The insights from the interviews will make it easier for you to figure out how to solve the problem. But remember that the information you get will only help if you know what you want to do. Set ambitious goals and use data to guide you towards them. Data may not tell you exactly where to go, but it can help you weigh the pros and cons of different ways to achieve your goal.

Qualitative and quantitative data are deeply connected to the product strategy and guide all product roadmaps in the right direction. They are the yin and yang of product insights. The following diagram shows you how product strategy is a good mix of qualitative and quantitative insights:

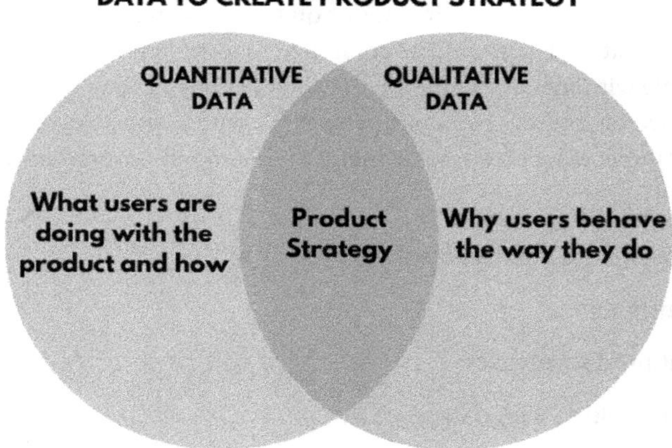

Figure 14.1 – Combining qualitative and quantitative data to create a product strategy

Using both qualitative and quantitative metrics can help remove blind spots and bias from the process of setting KPIs because it allows for a more comprehensive and diverse range of data to be considered.

Imagine you are developing a new API for a mobile app, and you have set a KPI to increase the number of API calls by 10% in the next quarter. To achieve this goal, you analyze quantitative metrics such as response time and error rates to identify any issues that may be affecting the API's performance.

However, you also gather qualitative data through customer feedback and usability testing. Through this process, you discover that many users were frustrated with the API's authentication process and found it difficult to use. This information may be yet to be captured through the quantitative metrics, but it is important to understand the API's overall performance and user experience.

By considering both quantitative and qualitative data, you can identify and address the issues with the authentication process, which ultimately leads to an increase in API calls and helps you achieve your goal. You need to consider both types of data to reach your goal. So, using both qualitative and quantitative metrics helps to remove blind spots and bias from the process of setting KPIs and ensures that the API is meeting the needs of its users.

Validating your insights

Quantitative data is easier to measure and can be tracked for change more frequently than qualitative data. You can use the various research techniques you learned in *Chapter 8* to gather qualitative and quantitative data. Since analyzing unstructured data from feedback forms, questionnaires, and diary studies that are part of qualitative data requires intense effort in conducting and interpreting, it is common to have less frequent updates to these insights. But qualitative data can give you more nuanced insights into the end user experience that can help you identify and creatively address the users' needs. Often, qualitative data can help you find opportunities you may not have been aware of yourself.

You can utilize qualitative data to drive decisions when you don't have enough data to make a decision. This is very useful when you are testing new functionality and want to get early feedback from customers. Usability studies are the best example of using qualitative data to shape the product journey. You can get low-fidelity mockups in front of a limited set of customers to get feedback. This allows you to get a signal from the customer even before you have quantitative data from exposing new functionality to a larger audience.

In scenarios where you identify patterns in quantitative data, such as a trend of declining usage of a particular feature, you can use qualitative data to validate and investigate this decline. The free-form feedback format in qualitative data will give you deeper insights into user decision-making as to why they are reducing their usage of said feature.

Defining the right product metrics

The importance of prioritizing certain product metrics over others is to narrow the focus and ensure that the product team focuses on the most important and relevant metrics. This helps avoid being overwhelmed by data and ensures that the team makes informed decisions based on the most impactful metrics.

In his book, *The Lean Startup* (`https://www.amazon.com/Lean-Startup-Entrepreneurs-Continuous-Innovation/dp/0307887898`), Eric Ries shares three criteria for good product metrics, which are shown in the following figure:

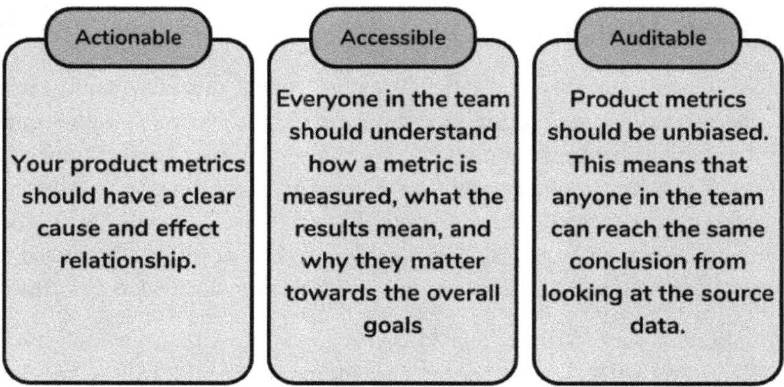

ERIC RIES' THREE CRITERIA FOR GOOD PRODUCT METRICS
Actionable
Your product metrics should have a clear cause and effect relationship.
Accessible
Everyone in the team should understand how a metric is measured, what the results mean, and why they matter towards the overall goals
Auditable
Product metrics should be unbiased. This means that anyone in the team can reach the same conclusion from looking at the source data.

Figure 14.2 – Eric Ries' three criteria for good product metrics

These criteria can also be applied to APIs to ensure that the metrics used are useful, relevant, and aligned with the overall business strategy:

- **Actionability**: Good product metrics for APIs should inform and guide the product team's decision-making process. For example, data on API response times and error rates can help the team identify and address performance issues. In contrast, API usage and adoption data can help inform marketing and growth decisions.
- **Accessibility**: Product metrics for APIs should be easily accessible to the entire product team. This means that the data should be readily available and not require significant effort to gather or analyze. For example, API analytics platforms can help track and monitor API performance in real time.
- **Auditability**: Good product metrics for APIs should align with the overall business strategy and goals. For example, if the company's goal is to increase revenue through the API, metrics such as the revenue generated or the number of paid API subscribers can help to track progress towards that goal.

Considering these three criteria, you can ensure that your product metrics for APIs are useful, relevant, and aligned with your business goals. This helps to inform and guide the API development process and ensure that the API is meeting the needs of its users.

If your product metrics meet all these conditions, you can be sure you're looking at the right data. In this case, the right data is the information that lets you keep track of the product strategy and goals and make decisions about them. On a more detailed level, it helps you decide which strategic themes to focus on when planning the product roadmap.

Another framework to evaluate your metrics for effectiveness is leading and lagging indicators, which you will learn about in the next section.

Leading and lagging indicators

Leading and lagging indicators are types of product metrics that can help track the performance and progress of a product.

Leading indicators are metrics that are predictive of future outcomes. These metrics are usually forward-looking and can help identify potential issues or opportunities before they occur. For example, a leading indicator for an API might be the number of new developers signing up to use the API, as this could be an indicator of future growth.

Lagging indicators, on the other hand, are metrics that reflect past performance. These metrics are usually backward-looking and can help to understand the results of past actions or events. For example, a lagging indicator for an API might be the total number of API calls made over a certain period, as this reflects the API's usage in the past.

Key characteristics of leading and lagging indicators can be summed up as shown in the following diagram:

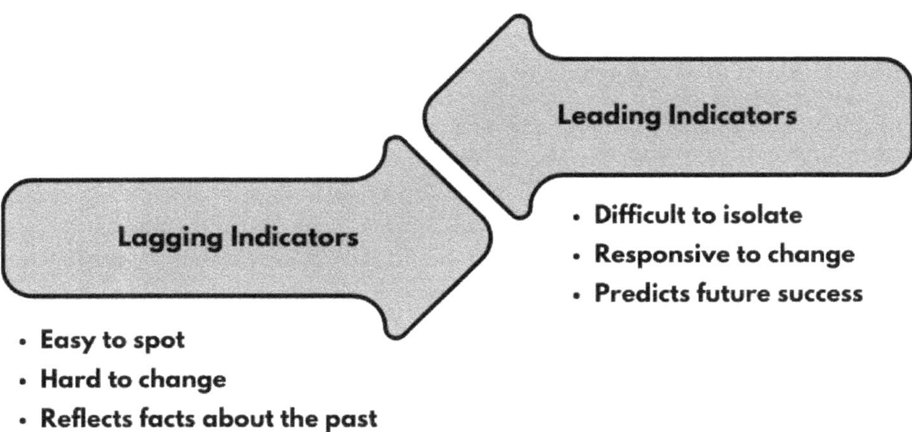
Leading Indicators
• Difficult to isolate
• Responsive to change
• Predicts future success
Lagging Indicators
• Easy to spot
• Hard to change
• Reflects facts about the past

Figure 14.3 – Differences between leading and lagging indicators

Both leading and lagging indicators are important for tracking the performance of a product and can help inform decision-making and identify areas for improvement. However, it is important to consider both types of metrics in order to get a comprehensive understanding of your product's performance and direction.

The challenge with lagging indicators is that they are metrics that reflect past performance, so they may take some time to materialize. This can make them less useful for teams that are trying to find out whether their current projects are working or not, as they may not provide timely feedback.

For example, if your team is working on a new feature for an API, it may take some time before the data on the number of API calls or revenue generated from the feature becomes available. This data would be considered a lagging indicator, as it reflects the results of your team's efforts in the past.

In contrast, leading indicators are metrics that are predictive of future outcomes and can provide timely feedback on the success or failure of a project. For example, if your team is working on improving the usability of the API, you may track metrics such as user satisfaction or the number of users completing specific tasks. These metrics can provide real-time feedback on the effectiveness of your team's efforts and allow them to make adjustments as needed.

Overall, while lagging indicators can be helpful in understanding the results of past efforts, there may be better ways for you to find out whether your current projects are working or not. Leading indicators, on the other hand, can provide more timely feedback and help teams make informed decisions about their current projects.

If you use a leading indicator as a stand-in for a lagging indicator, keep in mind that the leading metric should affect the lagging metric, but it might not have the effect you expect. Keeping track of both leading and lagging indicators so you can see how things are going and how your changes are working is key to getting a balanced view of the data.

Input and output metrics

Input metrics represent the actions you take, and outcome metrics represent the outcomes or results of your actions. Output metrics are the final outcome or goal you are trying to achieve. Outcome metrics form your North Star metrics, and these could be business metrics such as revenue, the number of active developers, and the number of transactions. These metrics are directly reflective of how the business is doing. It's important to understand that output metrics are driven by a number of input metrics, and you may not be able to directly influence them.

On the other hand, input metrics are actions within your control that you can take to influence output metrics. Effective input metrics are actionable and serve as goalposts and milestones to measure your efforts. Input metrics for a developer relations team might be to produce blog and video content that walks customers through the developer onboarding process for an API. But this input metric may or may not directly improve the output metric of reducing time-to-first-call across developers. You can directly control the content you produce to help users, but you cannot directly influence how the users consume this content or whether they get to make their first API call.

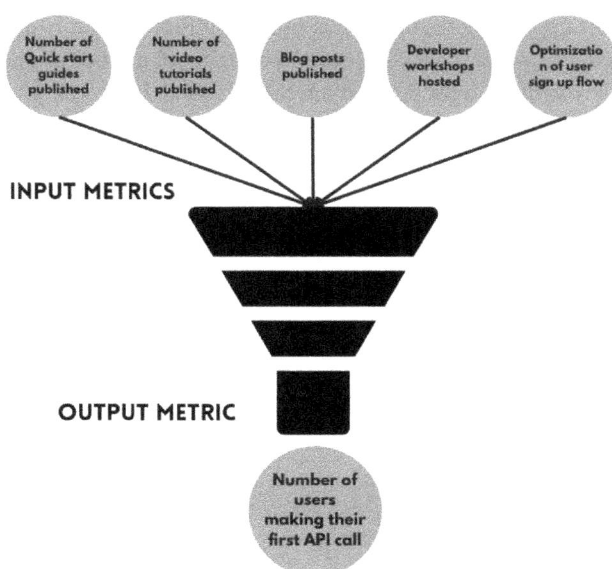

Figure 14.4 – Example of input and output metrics

Choose an input metric that you can control and make it actionable. You should pick an output metric based on the results you want. If you only set an output metric, the team will lose motivation quickly and feel like they are shooting in the dark because they can't see how their work affects the final output metric.

In the next section, you will learn about putting together the various types of analytics and metrics into a narrative and telling stories with data.

Framework for storytelling with data

Data by itself doesn't help you very much. You can gather as much of it as you want, but it won't bring people together or get them to do anything. Because data has so much to offer, when you look at your data, figure out what it means, and explain what you've learned in a way that everyone can understand, you've turned it into something very valuable: a story.

Stories are more interesting than facts because they stay with us longer. The same goes for giving information to your team and to the people in charge. Data storytelling can help people act on the things they learn from data. Without good communication, your audience might not notice or remember your insights. Hard and soft skills together help you get the most out of your data.

Telling stories is an important part of a product manager's job that is often overlooked. If someone asks you, as a product manager, what makes your product different and better than the competition, you can give a long list of reasons to back up your success. It's what it can do. It's the price, how well it works, how new it is, and so on. And you're probably telling potential customers the same things when you talk to them.

If you present your data as a bunch of unrelated charts and graphs, your audience might find it hard to understand or, even worse, come to the wrong conclusions. This can cause you to make bad decisions, which can hurt your business in a big way.

Here is a five-step framework that you can use to structure the storytelling exercise:

1. Identify the audience.
2. Develop a narrative.
3. Choose the right data and visualizations.
4. Draw attention to key information.
5. Engage your audience.

We will now dive deeper into each of these aspects.

Identify the audience

Before you start telling your story, you should first think about who you're going to tell it to. Think about what makes them tick, what they're interested in, and how to connect with them best. To win over an audience, you have to understand where they are coming from and what their priorities are, and then connect with them on an emotional and personal level.

When thinking about your audience, keep in mind that different people on your team have different goals and points of view. A good story based on data should talk about these differences. One good place to start is to think about how much your audience knows about the subject you're talking about.

Beginners may not know much about the subject but want to learn more. The bottom line is important to executives, and key statistics and KPIs help them decide what to do. You should pay attention to the most important lessons and how they affect the business as a whole. Business managers are mostly interested in themselves and want to make things better. You'll have to show them how your ideas can lead to results that people can see and touch.

Analysts are interested in the details and want to know how you do things. You'll have to show them how good your analysis is. Generalists care most about big ideas and big-picture analysis. Experts want information that is less based on stories and more based on questions. Supervisors want both details and insights that they can use. Executives are often in a hurry, so they want to know what the conclusions and implications are right away.

Develop a narrative

If you just show your data without explaining what it means, your audience will (at best!) look at it for a few seconds and move on, not really remembering any of the insights you've shared. A simple way of creating a narrative using different types of analysis is by answering the following questions in the following structure:

- *What happened?* Describe what happened using descriptive analytics. This paints a picture of the problem that you are trying to get the audience's attention to, and this gives you a chance to emphasize the size of the impact your solution could have.
- *Why did it happen?* Once you have painted the picture of the problem, you should share insights into the cause of the problem that helped you identify the possible impact and solutions. The type of analysis that helps you diagnose a problem and understand its causes is called **diagnostic analysis**.
- *What will happen?* Use predictive analysis to show what happens if the problem is left unsolved. Answering these questions helps the audience understand the risk of the problem.
- *How do we make good things happen?* Having identified the problem and justified the size of the impact, you can now come to the proposed solution on how you propose to fix it. You can use prescriptive analysis to show how you evaluated different factors and determined the next steps.

All four types of analysis—descriptive, diagnostic, predictive, and prescriptive—help you create a complete narrative that walks your audience through the depth of your analysis.

Choose the right data and visualizations

Data visualization is not the same thing as storytelling. Visualizations are still a critical part of creating a compelling narrative. When done right, data visualizations help people compare and understand information and put stories in the right context.

How do you create great data visualizations?

1. Pick the right data to show.
2. Choose the best way to show your data.
3. Make sure your visualization shows what's most important.

Find the parts of the data that show the exact points you want to make. Take out any information that isn't necessary to your story. If you give too much information, it's hard for readers to see the insights you want them to see. Use metrics and naming conventions that your audience will understand, such as *developer journey*, *time-to-first-call*, and so on.

Now that you know what to show, you can start creating the visualizations. Start by asking yourself what you want to get out of the visualization since each one has its own strengths and weaknesses. The following diagram shows you a simple way to choose chart types based on the nature of the information you are trying to present:

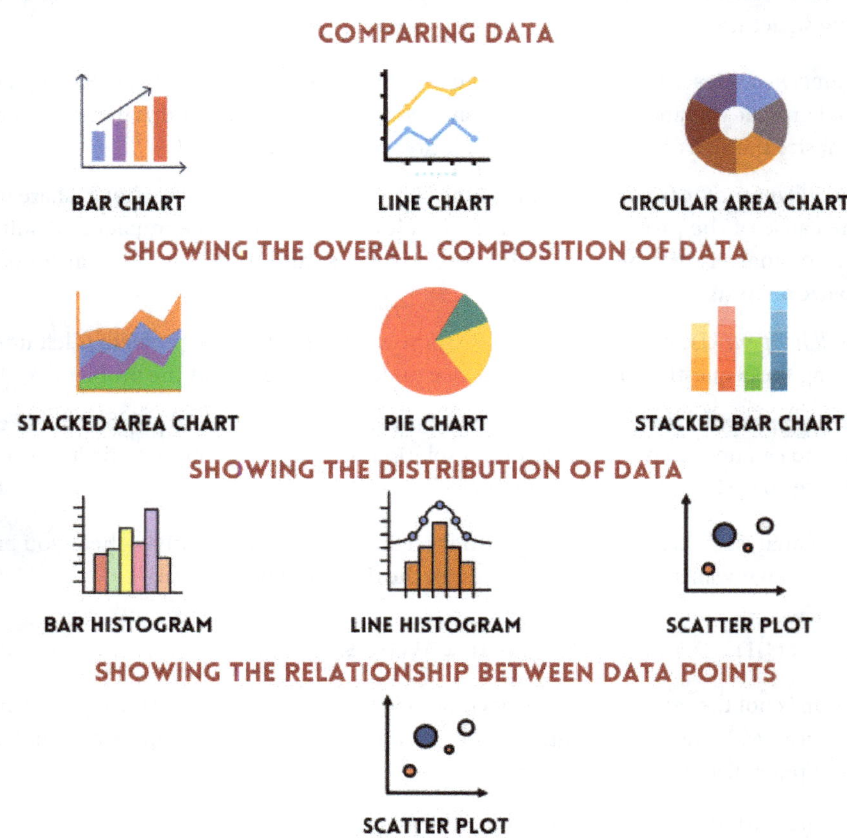

Figure 14.5 – Types of visualizations for different use cases

Avoid distracting styles such as shading, outlines, and 3D shapes that make your data less clear. Color can be a powerful tool for visualizations, giving clear clues about relative value, points of emphasis, and differences between datasets.

Draw attention to key information

If your graphs are complicated and don't highlight what's important, your audience might have trouble understanding what your story is about. They might even think the wrong thing. To tell a good story with data, you need to draw your audience's attention to the most important parts of your data. You can't expect your audience to know where to look or what to think—you have to show them. A few simple design tricks can help you get people's attention where it needs to be most.

Use color and titles to ensure all visualizations are easy to understand and are placed where they will draw the audience's attention. Size, titles, and color are keys to capturing the audience's attention and making a visualization easy to grasp.

Engage your audience

Stories temporarily get past our logical filter and reach us on an emotional level, which is where real decisions are made. When we are interested in a story, we put our doubts aside for a short time and listen with less skepticism. So, stories help us keep our minds on the big picture instead of the small details. This is a clear benefit for people who tell stories with data. And if you have a thorough analysis to back up your story, the impact can be even greater.

Summary

Having learned about the metrics frameworks in *Chapter 13*, you know how to actively find a balance of leading and lagging indicators in your metrics so that you can establish the right input and output metrics that your team can actively work on to have the desired impact on business goals. Having a balanced view of the data and being able to pick the right metrics for each initiative that your team works on is key to keeping your efforts focused and prioritizing the right items on your product roadmap.

Being able to narrate the strategy behind the metrics you choose and the changes in those metrics over time enables you to tell a compelling story to your stakeholders and leaders to gain their trust and support for your product initiatives. Now that you have learned how to tell stories with data, you will learn about identifying biases in your interpretation of data and eliminating blind spots in your metrics in the following chapter.

15
Counter Metrics to Avoid Blind Spots

In *Chapter 14, Keeping Metrics Honest*, you learned about establishing metrics in clusters. When you try to establish a vast set of balanced metrics, it is possible that you create biases or blind spots that you may not be aware of. In this chapter, you'll learn how to deal with biases and blind spots in the metrics you set up with the following strategies:

- Establishing counter metrics
- Avoiding gameable metrics
- Avoiding cannibalizing metrics
- Aligning incentives
- Avoiding cognitive biases

By being aware of your subconscious biases and the gameability of metrics, you can be on the lookout for these when setting your metrics.

Establishing counter metrics

Metrics-based product development is one of the best things to happen in the software age. Metrics let us define what success looks like and then track our progress until we've definitely reached our goals.

Metrics for product development come in a lot of different shapes and sizes, and each one has a different purpose. It is natural to pay the greatest amount of attention to the metrics that show how successful a product is. We use these **success metrics** or **goal metrics** to decide which features and improvements to add to our products first. They are quantitative measures of what we think success looks like, based on data. This makes it easy to measure success. They are the main way we know we are going in the right direction, and they are essential in measuring our operational efficiency.

For a given product, we might have a lot of metrics that show how well it's doing, or we might decide to focus on a single **North Star metric** for the most clarity and focus all of our efforts on that.

But overemphasizing one or two success metrics can have unintended effects on a business. This is where counter-metrics step in, to ensure that you are not blindsided by a focus on a handful of success metrics.

Counter metrics, also called **guardrail metrics** or **health metrics**, are measurements of parts of a product or business that may change if you focus on your success metrics. They are an important part of defining your success metrics because they help you see what's going on and stop things from going horribly wrong.

So, for every success metric, you should have one (or more) counter-metrics that serve as a constant reality check. Luckily for you, figuring out your counter metrics is as easy as thinking creatively about what might happen if you focus on your success metrics.

In a developer portal experience, you can create a success metric for the number of accounts created. This metric, although important to track, can be gamed if you start incentivizing users to sign up with different email addresses, in which case you are not truly acquiring new users. This metric can also be a vanity metric if users are signing up and creating an account but are not able to make their first API call. This is why it's important to look at the number of accounts created in the context of the number of users who make their first call. You can look at this data in the form of cohort analysis, or you can also track the rate of users reaching the first call by channel to understand the quality of users you're getting on the platform.

Still, it's important to know that counter-metrics need to be weighed against your business's top priorities. It might be smart to start by focusing on a small group of your possible users before moving on to the rest. So, when you're choosing which counter-metrics to use, you should think about any possible side effects and ask yourself how they could affect your goals.

Here are some examples of metrics and counter metrics for APIs:

Success metric	**Counter metric**
The number of API requests: This metric measures the number of requests that are being made to the API	**Latency**: This counter metric measures the amount of time it takes for the API to process a request and send a response
Success rate: This metric measures the percentage of successful responses to API requests	**Error rate**: This counter metric measures the percentage of failed responses to API requests
Response time: This metric measures the amount of time it takes for the API to respond to a request	**Throughput**: This counter metric measures the number of requests that the API can handle within a specific time frame

Table 15.1 – Examples of metrics and counter metrics

By keeping an eye on these metrics and counter metrics, you can find small problems with an API and fix them before they become big problems. This can help make sure that the API works well for users and gives them a good experience.

Avoiding gameable metrics

Gameable metrics are metrics that can be artificially manipulated, or **gamed**, to achieve a desired outcome. For example, if a metric is being used to track the success of an API product and that metric can be easily influenced by the actions of the team, it may be a gameable metric.

To avoid setting gameable metrics for an API product, it's important to choose metrics that are objective and difficult to manipulate. Here are some tips for avoiding gameable metrics:

- Use multiple metrics to track the success of the API product. This can help to ensure that the success of the product is not being measured by just one potentially gameable metric. Make sure the metrics you choose are tied to the overall goals and objectives of the API product. This will help to ensure that the metrics are measuring something that is important and relevant to the success of the product.
- Choose metrics that are based on objective data. For example, instead of using a metric that is based on subjective opinions, choose a metric that is based on things such as the number of API requests, the success rate of those requests, or the response time of the API.
- Avoid metrics that can be easily influenced by the actions of the team. For example, if a metric can be manipulated by the team simply by changing the way they use the API, it may be a gameable metric.

One example of a gameable metric for an API product could be the number of API calls made by developers. In the case of a payment API, if the goal of the product is to get more transactions to be processed, but the metric being tracked is the number of API calls, developers may be incentivized to design the API in a way that takes more API calls per transaction. This would not be an accurate reflection of actual usage and could cause the product to perform poorly, because the metric doesn't show how the API is really being used.

If developers are artificially inflating the number of API calls, it could mean that the product is being overused, which could potentially lead to server overload, slow response times, and other technical issues. Additionally, it could also lead to billing issues if the API usage is tracked and charged based on the number of calls made. Furthermore, if the product team is basing their decisions on this metric and the API is being heavily used, they may not be able to identify the true issues and opportunities related to the API's usage. This could lead them to make decisions that don't match up with what their users really want, which could lead to a bad product.

By following these tips, you can help ensure that the metrics you set for your API product are not gameable and are measuring the true success of the product.

Avoiding cannibalizing metrics

Metrics that measure the success of one part of a product or service at the cost of another are called **cannibalizing metrics**. For example, a cannibalizing metric could be a metric for an API product that counts the number of API requests but doesn't care about the quality of the responses.

To avoid setting cannibalizing metrics, you should make sure that a metric is tied to the overall goals and objectives of a product. This will help to ensure that the metric is measuring something important and relevant to the success of the product. You should also consider the trade-offs that may be involved in achieving the metric.

Using multiple metrics to track the success of the product helps ensure that the success of the product is not being measured by just one potentially cannibalizing metric. Involve all relevant stakeholders in the process of setting metrics. This can help to ensure that the metrics take into account the needs and priorities of all stakeholders and do not negatively impact other aspects of the product.

In addition to avoiding cannibalizing metrics, you can also align incentives across stakeholders to ensure that all the teams are working toward a common goal.

Aligning incentives

Alignment and accountability are important in goal setting for a product because they help to ensure that everyone involved in the product is working towards the same goal, and they are held responsible for meeting those goals.

When goals are aligned, it means that everyone understands what the goals are and how they fit into the overall strategy for the product. This can help to ensure that everyone is working toward the same objectives and that their efforts are focused in the right direction.

Accountability, on the other hand, helps to ensure that everyone is held responsible for meeting the goals that have been set. This can be achieved through regular check-ins and progress reviews, as well as through the use of metrics and **key performance indicators** (**KPIs**) to track progress toward the goals.

The purpose of aligning incentives within a team is to ensure that everyone is working toward the same goals and objectives and that their efforts are focused in the right direction. When incentives are aligned, it means that everyone on the team understands what the goals are and how their work fits into the bigger picture. This can help to ensure that everyone is working toward the same objectives and that their efforts are focused on achieving those goals.

Aligning incentives can also help foster a sense of teamwork and collaboration within the team. When everyone is working towards the same objectives and supporting one another in achieving those goals, it can create a sense of shared purpose and improve team cohesion.

There are several ways to align incentives within a team:

- Make sure that everyone on the team understands the overall goals and objectives of the product or project. This will help to ensure that everyone is working toward the same objectives and that their efforts are focused in the right direction.
- Clearly communicate how individual contributions fit into the overall goals and objectives of a product or project. This can help to ensure that everyone understands how their work fits into the bigger picture and how it is contributing to the success of the team.
- Offer incentives that are tied to the achievement of team goals. This can help to ensure that everyone is working toward the same objectives and that their efforts are focused on achieving those goals.
- Use metrics and KPIs to track progress toward the goals and objectives. This can help to ensure that everyone is held accountable for meeting the goals and that progress is being made toward them.
- Foster a culture of collaboration and teamwork. Encourage team members to work together and support one another in achieving the goals and objectives of the product or project.

By following these tips, you can align incentives within a team and help to ensure that everyone is working toward the same objectives and that progress is being made toward those goals.

In addition to being aware of the incentives of stakeholders and avoiding gameable metrics, you should also be aware of subconscious biases that might lead to picking incorrect or misleading metrics. You will learn about the types of biases in the next section.

Avoiding cognitive biases

Cognitive biases are mental shortcuts that people use to make sense of the world around them. These biases can influence how people perceive and interpret information and can lead to inaccurate or distorted judgments. There are many different types of cognitive biases, and they can affect various aspects of decision-making.

In the context of setting product analytics, cognitive biases can play a role in how metrics and goals are chosen and how progress toward those goals is evaluated. For example, if a team is setting goals for a product and they are affected by confirmation bias, they may only consider information that supports their preconceived ideas about the product and ignore information that contradicts those ideas. This can lead to the selection of metrics that are not representative of the true success of the product and may not accurately reflect the needs of the users.

To avoid the influence of cognitive biases in setting product analytics, it's important to be aware of these biases and take steps to mitigate their impact. This can include involving a diverse group of stakeholders in the process of setting metrics and goals, seeking out alternative viewpoints and perspectives, and using objective data to inform decision-making.

Being aware of cognitive biases can help product managers avoid making decisions that are based on distorted or incomplete information. It can also help them to be more open to alternative viewpoints and perspectives and to consider a broader range of options when making decisions. Although there are over a hundred different types of cognitive biases, in the following section, you will learn about the most common ones that you should keep in mind when setting and interpreting metrics across product usage, customer feedback, revenue and so on.

Confirmation bias

Confirmation bias is the tendency to look for, understand, and choose the information that backs up what you already think or believe. This bias can affect API metrics by making it more likely that metrics are chosen that don't show the real success of the API and may not accurately show what users want. The following diagram shows how our beliefs may not always overlap with data, and this leads to us ignoring evidence that doesn't agree with our beliefs, resulting in confirmation bias.

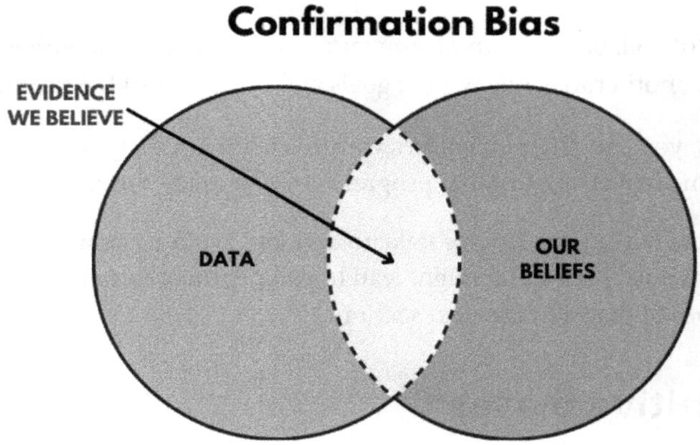

Figure 15.1 – Data being included and ignored in a case of confirmation bias

For example, if a team is affected by confirmation bias while setting metrics for an API, they may only look at the information that backs up what they already think about it and ignore information that goes against what they already think. This can lead to a selection of metrics that are not representative of the true performance of the API, such as those that are based on subjective opinions rather than objective data.

It is essential to be aware of confirmation bias and to take steps to reduce its effects if you want to avoid it when setting metrics for APIs. This can be done by including a wide range of stakeholders in the process of setting metrics, looking for different points of view, and making decisions based on objective data. By doing this, it's possible to avoid the effects of confirmation bias and ensure that the metrics are used to show how well the API really works.

Selection bias

Selection bias refers to a type of error that occurs when the sample that is being analyzed is not representative of an entire population. This can occur when the sample is selected in a way that is not random, and it can lead to incorrect conclusions being drawn about the population.

In the context of setting and interpreting product metrics for APIs, selection bias can occur if the sample of API users that is being analyzed is not representative of the entire population of API users. For example, if the sample only includes users who have a high level of engagement with the API, the metrics for that sample may be very different from the metrics for the entire population of API users. This can make it difficult to interpret the product metrics accurately and make informed decisions about the API based on those metrics. The following diagram shows how the selection of data used for analysis can lead to selection bias.

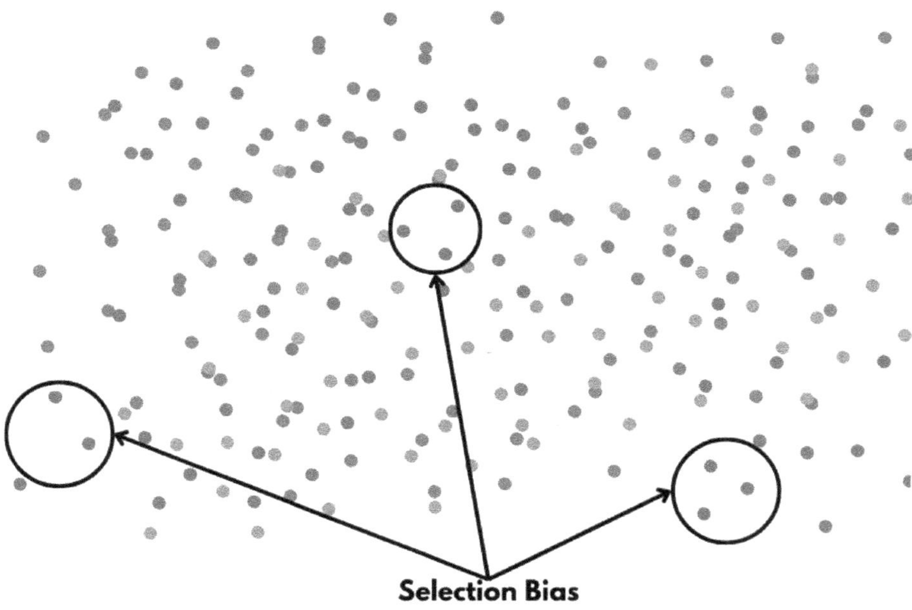

Figure 15.2 – Selection bias skewing our observation, based on the selection of data used for analysis

To avoid selection bias in this context, it is important to ensure that the sample of API users being analyzed is representative of the entire population. This can be done by selecting a random sample of API users or by using stratified sampling techniques to ensure that the sample is representative of different segments of the population. It is also important to consider any potential sources of bias that may be present in the sample and to take steps to mitigate those biases, in order to ensure the accuracy of the product metrics.

Anchoring bias

Anchoring bias is the tendency to rely too heavily on the first piece of information encountered when making decisions, even if that information is not relevant or reliable.

Imagine that you are a product manager at a software company, and you are analyzing the relationship between API usage and customer satisfaction. You collect a sample of data and plot the relationship between these two variables, and then you notice that there is a positive correlation – as API usage increases, customer satisfaction also increases.

However, as you are reviewing the data, you notice that one customer has a very high level of API usage and a very high level of satisfaction. This customer is an anchor, and you find that you are relying too heavily on this data point when making decisions about the relationship between API usage and customer satisfaction. You may be inclined to believe that the high level of API usage is the primary driver of the high level of customer satisfaction, even if other factors may be at play.

To mitigate the impact of anchoring bias in this scenario, it would be important to consider the full range of data and take a more holistic view of the relationship between API usage and customer satisfaction. This could involve looking at other factors that may be influencing customer satisfaction, such as the quality of the product or the level of support provided. By taking a more comprehensive approach, you can ensure that you are making informed, objective decisions based on data and evidence, rather than being influenced by a single data point.

Framing bias

Framing bias refers to the tendency to be influenced by how information is presented, rather than by the content of the information itself. The following diagram shows an image of two identical yogurts, one with a *20% fat* label and the other with an *80% fat-free label*. The framing of the labels biases how customers will perceive the two differently, even though they are identical.

Figure 15.3 – Framing bias demonstrated with yogurt labels

Imagine that you are a product manager at a software company, and you are conducting user research to understand how customers are using your product. You develop a survey to collect data on this topic, and you present the survey to a sample of customers.

However, as you are reviewing the results of the survey, you notice that how the questions were framed influenced the responses. For example, some questions were framed in a positive way ("How easy is it to integrate with our APIs?"), while others were framed in a negative way ("How difficult is it to integrate with our APIs?"). You find that customers are more likely to report that the product is easy to use when the question is framed positively and more likely to report that the product is difficult to use when the question is framed negatively.

To handle this framing bias, you could try reframing the questions in the survey in a neutral way ("How would you rate the ease of integration with our APIs?"), or you could use objective measures, such as log data or usage metrics, to assess product usage and usability. By avoiding framing bias, you can ensure that you are getting a more accurate and reliable picture of how customers are using your product.

Overconfidence bias

Overconfidence bias refers to the tendency to be more confident of your own judgment and decision-making ability than is warranted. In the context of analyzing product metrics, overconfidence bias can occur if you are too confident in your interpretation of data or unwilling to consider alternative explanations or viewpoints.

Imagine that you are analyzing the performance of your API and are interested in understanding the factors that influence the response time. You collect a sample of data and plot the relationship between different variables and the response time, and you notice that there is a strong negative correlation between the number of API requests and the response time – as the number of requests increases, the response time decreases.

Based on this data, you conclude that increasing the number of API requests will always result in a faster response time. However, upon further investigation, you realize that there are other factors, such as the complexity of the requests and the capacity of the API server. You find that the relationship between the number of requests and the response time is not as straightforward as you initially thought, and your initial conclusion was overly confident and not supported by the data.

To handle this overconfidence bias, you could try seeking out diverse perspectives, seeking feedback from others, considering multiple explanations for the data, and using objective measures to assess the performance of the API. By taking a more comprehensive and objective approach, you can ensure that you are drawing informed, reliable conclusions based on data and evidence.

Status quo bias

Status quo bias refers to the tendency to prefer the status quo and to be resistant to change. In the context of analyzing product metrics, status quo bias can occur if you are unwilling to consider alternative explanations or viewpoints, or if you are resistant to making changes based on data. The following diagram shows how status quo bias can lead to tunnel vision, where you explore solutions based on known variables, blinding you to solutions that lie outside of the status quo.

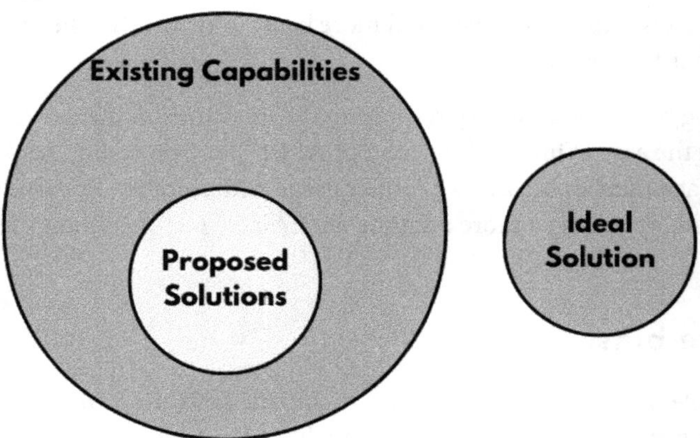

Figure 15.4 – Status quo bias limits your ability to explore solutions beyond the status quo

Imagine that a product manager is responsible for managing a software application that includes a GraphQL API that allows developers to query and modify data stored in the application. The API has been available for several years, and many developers have built integrations that rely on the existing API schema and functionality.

However, the product manager has received feedback from some developers that the API schema is difficult to understand and that certain parts of it are not well documented. They have suggested that adding more documentation and clarifying the schema could make the API more useful and easier to work with.

Despite this, the product manager is hesitant to change the API schema or add more documentation because they are concerned about the negative impact it might have on the existing integrations. They are worried that making changes to the schema or adding more documentation could cause confusion or disrupt the work of developers who are already using the API.

This hesitation to change the API is an example of status quo bias, as the product manager is inclined to maintain the current state of affairs rather than consider the potential benefits of making a change.

Outliers

Outliers are data points that are significantly different from the majority of data in a sample. When analyzing product metrics, outliers can have a significant impact on the results, as they can skew the mean and other statistical measures. There are several ways to handle outliers when analyzing product metrics:

- **Ignore them**: One option is simply to ignore the outliers and focus on the rest of the data. This may be appropriate if the outliers are not representative of the underlying population and are not expected to occur frequently.
- **Exclude them**: Another option is to exclude the outliers from the analysis. This can be done by setting a threshold for what is considered an outlier and excluding any data points that fall outside of that threshold.
- **Investigate them**: It is also important to investigate the outliers to understand why they are different from the rest of the data. This can help to identify any issues or problems that may be affecting the product metrics.
- **Use robust statistical methods**: Finally, it is possible to use statistical methods that are less sensitive to outliers, such as the median instead of the mean. This can help to reduce the impact of outliers on the analysis.

Imagine that you are analyzing the response time of your API and that you have collected a sample of 1,000 API requests. When you plot the response times, you notice that there is one data point that is significantly higher than the rest of the data. This data point could be an outlier.

You could ignore or exclude this data point completely, or you could investigate it further to understand its impact.

Recall bias

Recall bias refers to a type of error that occurs when individuals who are participating in a study have different rates of recall of past events or experiences. This can lead to incorrect conclusions being drawn about the relationships between different variables.

In the context of analyzing product metrics, recall bias can occur if the data being collected relies on individuals' recall of past events or experiences, and if there are differences in the accuracy or completeness of recall between different groups of individuals. Product managers partner with user research teams to evaluate customer interviews, and as you consume the insights from customer quotes, you can consider whether recall bias was introduced at any point.

Here are a few ways to handle recall bias when analyzing product metrics:

- **Use objective measures**: One way to reduce the risk of recall bias is to use objective measures whenever possible rather than relying on individuals' recall. For example, you could use log data or other types of automatically collected data to measure product usage or performance, rather than relying on individuals' recollections of their past experiences.
- **Use validated instruments**: Validated instruments, such as standardized questionnaires or surveys, can also be used to reduce the risk of recall bias. These instruments are designed to be consistent and reliable, and they can help to reduce the impact of differences in recall between different groups of individuals.
- **Use multiple sources of data**: Collecting data from multiple sources can also help to reduce the risk of recall bias. For example, you could combine data from log files, surveys, and interviews to get a more comprehensive picture of the product metrics.

Imagine that you are analyzing the usage of your API and are interested in understanding the factors that influence the number of API requests made by different users. You decide to conduct a survey to collect data on this topic, and you ask users to recall the number of API requests they made over the past month.

However, you find that some users have much higher recall accuracy than others, and this is affecting the results of your analysis. For example, users who are highly engaged with the API may be more likely to accurately recall the number of API requests they made, while users who are less engaged may be less accurate in their recall. This could lead to incorrect conclusions being drawn about the factors that influence API usage.

To handle this recall bias, you could consider using objective measures of API usage, such as log data, or you could use validated instruments to collect data on API usage more consistently and reliably. You could also consider collecting data from multiple sources, such as surveys and log files, to get a more comprehensive picture of API usage.

Confounding bias

Confounding bias occurs when the relationship between two variables is distorted by the presence of a third variable. In the context of analyzing product metrics, confounding bias can occur when the relationship between the metric being measured and the factor being studied is distorted by the presence of other variables.

Here are a few ways to handle confounding bias when analyzing product metrics:

- **Control of confounding variables**: One way to reduce the impact of confounding bias is to account for the confounding variables in the analysis. This can be done by stratifying the sample based on the confounding variable or by adjusting for the confounding variable in the statistical analysis.
- **Use experimental designs**: Experimental designs, such as randomized controlled trials, can be used to help control the effect of confounding variables by randomly assigning subjects to different groups. This can help to ensure that the groups being compared are as similar as possible, reducing the impact of confounding variables on the results.
- **Use multivariate analysis**: Multivariate analysis techniques, such as regression analysis, can be used to account for multiple confounding variables occurring simultaneously. This can help us to understand the relationships between the variables better and reduce the impact of confounding bias on the results.
- **Consider the limitations**: It is essential to be aware of the potential for confounding bias when analyzing product metrics and to consider the limitations of the analysis, in the context of any confounding variables that may be present.

Imagine that you are analyzing the performance of your API and are interested in understanding the relationship between the number of API requests and the response time. You collect a sample of data and plot the relationship between these two variables, and you notice a positive correlation – as the number of requests increases, the response time also increases.

However, upon further investigation, you realize that there is a third variable that is confounding the relationship between the number of requests and the response time – the complexity of the request. You find that the more complex the request, the longer the response time, and you also find that the number of requests is correlated with the complexity of the request. As a result, the apparent relationship between the number of requests and the response time is actually being distorted by the complexity of the request.

To handle this confounding bias, you could try controlling for the complexity of the request in your analysis. For example, you could stratify the sample by the complexity of the request and analyze the relationship between the number of requests and the response time separately for each stratum. Alternatively, you could use multivariate analysis techniques, such as regression analysis, to account for the complexity of the request and other potential confounding variables. This can help to better understand the true relationship between the number of requests and the response time, thereby reducing the impact of confounding bias on the results.

Association bias

Association bias refers to a type of error that occurs when the relationship between two variables is distorted by the presence of a third variable. In the context of analyzing product metrics, association bias can occur if the relationship between the metric being measured and the factor being studied is distorted by the presence of other variables.

Here are a few ways to handle association bias when analyzing product metrics:

- **Control of confounding variables**: One way to reduce the impact of association bias is to control for the confounding variables in the analysis. This can be done by stratifying the sample based on the confounding variable, or by adjusting for the confounding variable in the statistical analysis.
- **Use experimental designs**: Experimental designs, such as randomized controlled trials, can be used to help account for confounding variables by randomly assigning subjects to different groups. This can help to ensure that the groups being compared are as similar as possible, reducing the impact of confounding variables on the results.
- **Use multivariate analysis**: Multivariate analysis techniques, such as regression analysis, can be used to account for multiple confounding variables at the same time. This can help to understand the relationships between the variables better and to reduce the impact of association bias on the results.
- **Consider the limitations**: It is important to be aware of the potential for association bias when analyzing product metrics and to consider the limitations of the analysis, in the context of any confounding variables that may be present.

Imagine that you are analyzing the relationship between API usage and customer satisfaction. You collect a sample of data and plot the relationship between these two variables, and you notice that there is a positive correlation – as API usage increases, customer satisfaction also increases.

However, upon further investigation, you realize that there is a third variable that is confounding the relationship between API usage and customer satisfaction – the paid support tier. You find that customers who pay for a higher level of support are more satisfied with the API, and you also find that API usage is correlated with the paid support tier. As a result, the apparent relationship between API usage and customer satisfaction is actually being distorted by the presence of the paid support offering.

To handle this association bias, you could try accounting for the support tiers in your analysis. For example, you could stratify the sample by the support tiers and analyze the relationship between API usage and customer satisfaction separately for each stratum. Alternatively, you could use multivariate analysis techniques, such as regression analysis, to account for the support tier and other potential confounding variables. This can help you to understand better the true relationship between API usage and customer satisfaction, thereby reducing the impact of association bias on the results.

Biased data can result in the misinterpretation of facts and decisions that might lead to undesirable results. However, not every type of bias is measurable, and it's possible that these biases creep into your analytics strategy. That is the nature of biases. Although you may not be able to completely avoid subconscious cognitive biases in your analysis, it's important to be aware of them and revisit them regularly to make sure you are not blindsided by your own biases.

Summary

In this chapter, you learned that by aligning the incentives of various stakeholders while avoiding gameable metrics, cannibalizing metrics, and cognitive biases, you can create a more successful analytics strategy for your product.

It is important to ensure that everyone involved in the product has a clear understanding of the goals and objectives of the product, as well as how their work contributes to those goals. This can help to ensure that stakeholders are aligned in their incentives and are working toward a common set of objectives, rather than pursuing their own agendas.

You learned how to avoid gameable metrics by choosing metrics that accurately reflect the desired outcomes and cannot be easily gamed or manipulated. You also learned how to avoid cannibalizing metrics by ensuring that the metrics being used to track the success of the API product do not conflict with or undermine the metrics being used to track the success of other products or initiatives.

Finally, by being aware of cognitive biases, you can be mindful of how your own beliefs, preconceptions, and mental shortcuts can influence your interpretation of data and your decision-making processes. By being aware of these biases and taking steps to mitigate their impact, product managers can make more objective and data-driven decisions.

As you think about keeping metrics honest and balanced to avoid blind spots and biases in your analytics strategy, it is also important to bring all these thoughtful decisions together and create a narrative that your team and leadership can consume. In the next chapter, you will learn about bringing together your insights from analytics and telling effective stories with data.

16
Decision-Making with Data

As a product manager, one of your key responsibilities is to make decisions about the product and its direction. Data can be a powerful tool to inform these decisions by providing insights about customer behavior, market trends, and the performance of the product. By using data to inform decision-making, you can make more informed, evidence-based decisions that are more likely to lead to the success of the product. This can involve analyzing data to identify opportunities for product improvements, developing hypotheses about how changes to the product will impact its performance, and using data to measure the success of these changes. Ultimately, the purpose of using data to inform decision-making is to make better, more informed decisions that will help the product achieve its goals and succeed in the market.

Using data to make decisions also ensures consistency in your decision-making and helps you gain alignment across stakeholders. Following the data trail is a repeatable process that can enable small as well as large teams to operate in unison and make sure product initiatives across the organization are complementary.

Once you develop an understanding of the customer journey, the API experience components, and the process of building and managing API products, you can implement an analytics strategy that enables you to create well-informed product roadmaps that will make your team and product successful. The techniques that you will learn about in this chapter include the following:

- Bringing it all together
- Short-term goals
- Long-term goals
- Strategic storytelling
- Leading the team to success

In this chapter, you will learn about how data helps you to put together short-term and long-term goals and be an effective leader using strategic storytelling.

Bringing it all together

Applying a data-driven approach to building products allows you to bring your team and stakeholders into alignment with the priorities you set and make you confident in the decisions you take. Some of the key benefits of a data-driven approach to decision-making are the following:

- **You'll be more certain of the decisions you make**: Once you start collecting and analyzing data, you'll probably find it easier to make confident decisions about almost any business challenge, whether you're deciding to launch or stop selling a product, change your marketing message, expand into a new market, or do something else.
- **Data performs multiple roles**: On the one hand, it lets you compare what you have to what already exists. This helps you understand how any decision you make will affect your product roadmap.

  Beyond this, data is logical, repeatable, and concrete in a way that gut instinct and intuition are not. By taking out the subjective parts of your business decisions, you can give yourself and your company more confidence. Because of this confidence, your organization can fully commit to a certain vision or strategy without worrying too much that the wrong choice was made.

  Even if a decision is based on facts, that doesn't mean it's always right. Even if the data shows a certain pattern or points to a certain outcome, any decision based on the data would be wrong if the way it was collected or interpreted was wrong. This is why the effects of every business decision should be measured and kept track of on a regular basis. Oftentimes, there is insufficient data and you are not able to reach statistical significance. In such scenarios, you can identify the need to capture more data before you can draw actionable insights.
- **You'll start to take more initiative**: When you first start using a process that is based on data, it is likely to be reactionary. Make sure to complement the data available to you internally with data you can get for external factors such as the changing market landscape. This data tells a story, which your product strategy must then respond to.

  Even though this is useful in and of itself, it's not the only thing that data and analysis can do for your business. With enough practice and the right kinds and amounts of data, you can use it more proactively, such as by finding business opportunities before your competitors do or spotting threats before they get too bad.
- **You'll be able to save money**: In addition to the many reasons why a business might decide to put money into a *big data* project and try to make its processes more data-driven, cost savings tend to be a prominent one. Making smart decisions based on the most accurate and up-to-date information is quickly becoming the norm.

Data-driven decision-making makes it much faster to find a threat or challenge and figure out how to deal with it. This makes it much less likely that the threat will hurt your business. Analytics can also help figure out how internal operations and processes work and find ways to improve them that might not have been seen otherwise.

Short-term goals

You should combine short-term analytics goals with a *strategic picture* of how analytics can be used to create business value. The organization's strategic analytics framework and roadmap should include this *strategic picture*. Short-term goals and projects in analytics are important because they do the following:

- Build and/or improve the organization's momentum as it starts or continues its analytics journey
- Validate the value of data analytics for the business
- Give you a chance to try out and quickly learn how analytics models and tools can be used in the organization's context
- Confirm the business case for investing in analytical tools and platforms

Short-term analytics goals and objectives shouldn't take more than two to three months to implement, depending on the size of the organization and how easy it is to get the data. They may also need to use both structured and unstructured data.

Structured data is information that has been put into a set field in a file or record. Structured datasets are things such as the data in Microsoft Excel spreadsheets or an Access database. Unstructured data, such as video, PDF documents, social media data, and emails, is not set up in a traditional database or spreadsheet format and may not be as easy to analyze (in comparison to structured data).

The majority of the data in the world is unstructured, and only a small portion of the data is structured. Some estimates say only about 20% of the world's data is structured, while the rest is unstructured. An effective data strategy needs to use both structured and unstructured data to help make better-quality decisions and add value to the business.

Short-term goals and objectives for analytics can help answer more immediate questions, such as the following:

- What are the most effective channels for driving customer growth?
- Which one of my products makes me the most money?
- How many people will my customer support team need next quarter?
- How many customer questions should I expect to receive next quarter?
- Who are my most important customers?
- What is the average lifetime value of a customer?
- Which of my products does social media talk about the most?
- How many people visit our websites every month, and where do they come from?

Short-term analytics goals should use as much existing (internal or external) data as possible to get things done quickly and give the organization quick *analytics wins*. One benefit of a shorter horizon is that planning for the near future is easier and more accurate. It's much easier to predict the near future and make plans than to plan things out a few years in advance.

Even if the dates are likely to change, they should be included in a short-term roadmap because they give a sense of scale and a starting point for each major release or milestone. Whether you show or hide the dates depends on whom you're talking to. It may make sense to show them internally. Offering dates to customers and prospects could be seen as a commitment instead of a goal, and you might risk making some customers unhappy if you don't ship when the roadmap says you would.

Long-term goals

A product's mission defines the purpose or overall goal of the product in the long term. It should make it clear what makes the product special, what problem it's trying to solve, and who it's meant for. A mission statement should be short and to the point, and it should be used to guide decision-making and ensure that all actions align with the product's overall purpose.

A product's vision, on the other hand, defines the desired future state of the product. It should be an ambitious goal that shows what the product's long-term goals are and what effect it wants to have on its audience. A vision statement should be inspirational and should be used to guide the overall strategy and direction of the product.

In summary, the mission statement answers the question, "What do we do?" and the vision statement answers "What do we want to achieve?".

The mission statement is the foundation of the product's strategy; it defines the reason for the product's existence, the problem it's solving, and the target audience it's serving. The vision statement is the long-term goal of the product; it defines where the product wants to be and what kind of impact it wants to make.

You should set long-term goals for your product because they provide a roadmap for the product's development and success. Long-term goals help to ensure that the product remains focused on its overall vision and mission rather than being distracted by short-term priorities or opportunities. By setting long-term goals, you can also ensure that the product is continuously evolving and improving over time. This can involve setting goals around KPIs such as user growth, revenue, or customer satisfaction, as well as goals related to the product's features and capabilities. In addition to helping to guide the product's development, long-term goals can also help to motivate and focus the product team and provide a sense of purpose and direction for the work that they do. Ultimately, long-term goals are an essential part of the product management process, as they help to ensure that the product remains focused on achieving its long-term objectives and delivering value to its customers.

Creating a roadmap for the next two or five years may seem like a waste of time in a world where decisions are made quickly and based on data. But in many ways, the things on a long-term roadmap are much more significant than what you plan to do in the next few quarters.

A long-term roadmap starts with the vision for the product and shows how to get there. It takes something that is just a theory and turns it into something that makes sense. It also gives a rough timeline for how big dreams and ideas can become reality.

Long-term roadmaps don't need to be changed and updated as often as short-term ones. If your vision changes more than twice a year, you probably have more significant problems than an old document. Instead, long-term roadmap updates should result from work on vision and strategy that leads to new findings, conclusions, or decisions made by the product team, executives, and stakeholders. A long-term roadmap can ensure that both stakeholders and contributors agree with the vision and strategy.

Since this roadmap doesn't change very often, a new version should probably have an official presentation and review explaining the changes being made and why they were made. It's a great chance to remind everyone in the company what the ultimate goal of the product is and how their current work and day-to-day tasks fit into that vision.

It should also be clear that the long-term roadmap is not a list of promises or deliverables from the product team. The future is too far away to assume that any one thing will happen, let alone when it is currently scheduled. The long-term roadmap concerns the big picture, not when something will come out.

Here are some dos and don'ts when it comes to long-term product goals:

- Do set ambitious but realistic goals that align with the overall vision of the product
- Do consider the long-term impact of decisions made in the short term
- Do regularly review progress and adjust goals as necessary to ensure they remain relevant
- Do involve all relevant stakeholders in the goal-setting process
- Don't neglect to consider the feasibility of long-term goals
- Don't forget to communicate long-term goals clearly and regularly to all stakeholders
- Don't forget to set intermediate goals that will help achieve long-term goals

Having learned about setting short- and long-term goals, you will learn about putting together a product strategy document in the next section.

Writing a product one-pager

A product strategy document is a document that outlines the goals and objectives for a product, as well as the strategies and tactics that will be used to achieve those goals. This document is also sometimes referred to as a *product one-pager* because the goal of this document is to communicate the strategy of the product in the most concise and succinct way possible. Most organizations have their own templates for writing product strategy documents, but these documents typically include information such as the target market, competitive landscape, and the key features and benefits of the product.

To write a product strategy document for an API product, you will need to consider the following elements:

- **Target market**: Identify the specific industries and companies that would benefit from using the API. Use market research to identify the market segments and audiences that would be the best fit for your proposed product. Use top-down, bottom-up, or theory of value strategies for calculating the **total addressable market (TAM)** for this section, which you learned about in *Chapter 5*.
- **Competitive landscape**: Analyze the existing API offerings on the market, and identify how your API differentiates itself. Use the **strengths, weaknesses, opportunities, and threats (SWOT)** analysis method that you learned in *Chapter 13* to evaluate your product's competitive position in the market. This is also your opportunity to highlight customer pain points that are currently unaddressed by existing products.
- **Value proposition**: Outline the key features and benefits of the API that will appeal to your target market and solve their specific problems. Dive into the details of how your solution delights customers and why customers would want to use your product.
- **Monetization strategy**: Determine how you will generate revenue from the API, such as through subscription fees or usage-based pricing. Having studied the competitive landscape and understanding the value that your product will deliver to customers, you can derive the right pricing strategy (*Chapter 5*) for your product offering.
- **Go-to-market strategy**: Outline the tactics that will be used to promote and sell the API, such as targeted marketing campaigns, partnerships, and developer outreach. Sales, marketing, and developer relations teams are your key stakeholders in establishing the go-to-market strategy. This is where the skills in stakeholder alignment that you learned in *Chapter 15* will come in handy.
- **Roadmap**: Define the development, launch, and update plans for the API and its features. Partner with your engineering leads to define the scope of the **minimum viable product (MVP)** and specify the included features. You can use the API maturity framework you learned about in *Chapter 3* to establish the different phases of your product development and establish the estimated timeline for all the phases of your product launch.

- **Metrics of success**: Define how you will measure the success of the API, such as user engagement, revenue, and adoption rate. Use the framework for identifying the infrastructure needs you learned about in *Chapter 10* to make sure you have the infrastructure scoped for your product. Use the product metrics and business metrics you learned about in *Chapter 11* and *Chapter 12*, respectively, to establish KPIs for various phases of your product development journey.

The structure of this document might vary from organization to organization. You can also add sections to capture information that might be specific to your product and domain. It's important to keep in mind that the product strategy document should be a living document and should be reviewed and updated regularly to reflect the changing market and customer needs. Let us now look at an example of a product strategy document using this framework.

Sample strategy document

Imagine that you are the product manager proposing to build a new "Smart Traffic API" and you want to present this product idea to the leadership in your organization. Here is an example of what your product one-pager might look like:

Target Market:

- Cities and municipalities looking to optimize traffic flow and reduce congestion
- Transportation companies looking to improve route planning and delivery efficiency

Competitive Landscape:

- There are several existing traffic APIs on the market, but most are limited in scope and do not offer the level of granularity and real-time data that our API will provide

Value Proposition:

- Our API provides real-time traffic data for major roads and highways, as well as detailed information on traffic incidents and road closures
- Our API also offers advanced features such as predictive traffic analysis and route optimization for transportation companies
- This will help cities and municipalities to make better-informed decisions about traffic management and transportation companies to optimize their routes, saving time and fuel costs

Monetization Strategy:

- We will offer a freemium model, where basic traffic data is available for free, but advanced features such as predictive analysis and route optimization will be offered on a subscription basis
- We will also offer a pay-per-use pricing model for high-volume users

Go-to-Market Strategy:

- We will target cities and municipalities through targeted online and offline advertising campaigns
- We will also reach out to transportation companies through industry events and online marketing campaigns
- We will also offer free trials and demo versions of the API to encourage usage and adoption

Roadmap:

- Phase 1 (Q1): Development of the basic traffic data API
- Phase 2 (Q2): Release of advanced features such as predictive traffic analysis and route optimization
- Phase 3 (Q3): Release of the API for international cities

You could also present this information in a visual roadmap highlighting every activity on a timeline as shown in the following figure:

Figure 16.1 - Roadmap showing three phases of the Smart Traffic API

Product roadmaps are helpful for all stakeholders to have a single artifact they can all refer to in order to plan their respective team's efforts. Ideally, your roadmap should show when features will be available for beta testing and when they will be available for general release. This way, sales and operations teams can plan their external-facing communication strategy.

Metrics of Success:

- Daily, weekly, and monthly new user sign-ups
- Time to first API call (TTFHW)
- Number of API calls per month
- Number of paying customers
- Average revenue per user
- Customer retention rate

This is just an example, and it should be tailored to the specific product and target market. Also, it should be kept updated as the market and customers change.

Strategic storytelling

Strategic storytelling is an important tool for product managers because it helps them communicate the value and vision of the product in a way that is compelling, inspiring, and motivating for the product team and other stakeholders. This can include telling the story of the product to potential customers, investors, partners, and even the product team itself. By using storytelling to communicate the product's value proposition and unique selling points, you can help build excitement and support for the product. You can also use storytelling to articulate the product's vision and direction and to paint a picture of what the future could look like if the product is successful.

In addition to helping to communicate the value of the product, strategic storytelling can also be an effective tool for shaping the product's direction. By telling a compelling story about the product, you can help align your team around a shared vision and inspire them to work toward a common goal. You can also use storytelling to get buy-in from key stakeholders, such as executives or investors, and persuade them to support the product and its development.

There are several methodologies that you can use to tell effective stories about your product:

- **The Hero's Journey**: This is a storytelling structure that follows the journey of a hero as they face challenges and overcome obstacles on the way to achieving their goal. This structure can be used to tell the story of the product and its journey to solve a problem for customers.
- **The Problem-Solution Narrative**: This is a storytelling structure that involves outlining the problem that the product is trying to solve, and then describing how the product provides a solution to that problem.
- **The Before-After Narrative**: This is a storytelling structure that involves describing the current state of the problem or challenge that the product is trying to solve, and then explaining how the product transforms the situation into a better state.

- **The Why-How-What Structure**: This is a storytelling structure that involves explaining the purpose or motivation behind the product (the Why), describing how the product works (the How), and then outlining the features and benefits of the product (the What).
- **The Past-Present-Future Structure**: This is a storytelling structure that involves describing the history or background of the product (the Past), explaining the current state of the product (the Present), and then outlining the future vision and goals for the product (the Future).

The most effective storytelling methodology for you will depend on the specific goals and objectives of your product, as well as the audience you are trying to reach. You can use data to tell effective stories by using it to support and illustrate your narrative. This can involve using data to highlight key points and make them more concrete and convincing. For example, if you are trying to communicate the value of a new feature, you could use data to show how it has improved key metrics such as customer retention or conversion rates. By using data to support your story, you can help to make your narrative more credible and persuasive.

In addition to using data to support your story, you can also use data to help identify the most compelling angles and themes to focus on. This can involve analyzing data to identify trends and patterns that can be used to illustrate the product's value proposition or unique selling points. For example, you could use data to identify the most common problems that your product solves for customers or to highlight the most popular features of your product.

Finally, you can use data to help visualize and communicate your story in a more impactful way. This can involve creating charts, graphs, and other visualizations that help to illustrate key points and make the story more accessible and engaging for the audience. By using data to create visually appealing and easy-to-understand graphics, you can help make your story more memorable and effective.

Imagine that you are responsible for a new API that allows developers to access and integrate data from a popular social media platform into their own apps. You have been noticing that adoption of the API has been slower than expected, and you want to understand why this is happening and how to increase usage.

To tell an effective story about the problem and the solution, you could use data in the following way:

1. **Identify the problem**: You could use data to identify the key reasons why developers are not using the API. This might involve analyzing usage patterns, tracking customer feedback, and looking at metrics such as API calls and integration rates.
2. **Describe the solution**: Once you have a clear understanding of the problem, you can use data to help describe the solution that the API provides. This might involve highlighting the key features and capabilities of the API, or showing how it has helped other developers to build more successful apps.

3. **Use data to support the story**: To make the story more convincing and impactful, you could use data to illustrate the value of the API. This might involve showing how the API has helped developers to increase their user engagement or revenue, or how it has allowed them to build more sophisticated and differentiated apps.

Overall, by using data to identify the problem, describe the solution, and support the story, you can create a compelling narrative that illustrates the value and benefits of the API and helps to convince developers to adopt it.

Leading the team to success

Product managers lead their teams to success by setting clear goals and objectives, developing a strong vision and strategy for the product, and effectively communicating with and motivating the team to work towards these goals. Some specific ways in which you can lead your teams to success include the following:

- **Setting clear goals and objectives**: You should set clear and measurable goals for the team to work toward and ensure that these goals are aligned with the overall vision and mission of the product. This can involve setting both short-term goals to help the team make progress on immediate priorities, as well as long-term goals to guide the product's development over time.
- **Developing a strong vision and strategy**: You should have a clear vision and strategy for the product and be able to articulate this vision to the team in a way that inspires and motivates them to work toward it. This can involve identifying the product's unique selling points, developing a roadmap for the product's evolution, and identifying KPIs to measure success.
- **Communicating effectively**: You should be a skilled communicator and be able to effectively communicate your vision and goals to the team, as well as listening to and incorporating feedback from team members. You should also be able to clearly articulate the roles and responsibilities of each team member, and provide guidance and support as needed.
- **Motivating and inspiring the team**: You should be able to inspire and motivate the team to work toward the goals and vision of the product. This can involve providing recognition and rewards for excellent performance, creating a positive and collaborative work environment, and helping team members to develop their skills and careers.

Overall, product managers lead their teams to success by setting clear goals and objectives, developing a strong vision and strategy, and effectively communicating and motivating the team to work toward these goals.

Summary

Through the chapters of this book, you learned about the role of the API product manager, API lifecycle, maturity models, customer journey, designing effective API experiences, infrastructure, business and product metrics for APIs. These frameworks would enable you to get the insights that you can use to set an effective product strategy for your APIs.

Putting together a product strategy can be challenging but bringing together market research and customer stories enables you to establish an inspiring mission and vision that your team would be excited to work towards. While well-defined long-term goals will help you communicate to your stakeholders how you plan to realize the vision for your product, short-term goals help various teams involved with your product to plan their work and find alignment as you work towards shared goals.

APIs are no different from any other products. You have a unique opportunity to bring customer empathy and data-driven decision-making to the rapidly evolving world of API products, and I hope this book has provided you with the structure and frameworks to do so!

The API Analytics Cheat Sheet

API metrics can be segmented into three main categories: Infrastructure, Product Experience, and Business metrics.

Infrastructure metrics provide insight into the performance and reliability of the API, such as uptime and availability. These metrics help ensure that the API is functioning correctly and that users are able to access it smoothly. Product Experience metrics measure how users interact with the API, such as the active users and time to first transaction. These metrics help understand how the API is being used, and identify areas for improvement. Business metrics focus on the impact of the API on the overall business, such as revenue, customer satisfaction, and user engagement. These metrics help understand how the API is contributing to the business objectives and goals.

For a good API analytics strategy, it is important not to try to set up all the metrics on Day 1. Instead, it is better to start by focusing on a few key metrics that align with the current objectives, and then expand and adjust the metrics as the API and the business evolve over time. Use the sheet below to identify the Day 1 and Day 2 metrics that you already have established for your product.

Infrastructure Metrics for APIs		
	Day 1	**Day 2**
Performance	☐ Uptime and availability	☐ Errors per minute ☐ Average and max latency ☐ 90th percentile latency by customer
Usage	☐ Request per minute (RPM) ☐ Top Endpoints ☐ Usage by segments	☐ CPU Usage ☐ Memory Usage ☐ Error code distribution ☐ Concurrent Connections
Reliability	☐ Mean Time to Failure (MTTF) ☐ Mean Time to Repair (MTTR) ☐ Mean Time Between Failure (MTBR)	☐ Rate of occurrence of failure (ROCOF) ☐ Probability of failure on Demand (POFOD)
Product Metrics for APIs		
	Day 1	**Day 2**
Discovery	☐ Unique Visitors ☐ Page Views ☐ Signups by Channel	☐ Reading Level or Text Complexity Analysis ☐ Link Validation
Engagement	☐ Bounce Rate ☐ Customer engagement score (CES)	☐ Average time on page ☐ Search Keyword Analysis ☐ Engagement with tools
Acquisition	☐ Daily User Signups - New User ☐ SDK and Version Adoption	☐ Time to First Hello World (TTFHW)
Activation	☐ Time to First Transaction (TFT) ☐ Active Users (DAU/WAU/MAU)	☐ Time to value (TTV) ☐ Cohort Analysis
Retention	☐ Recurring D/W/M Usage	☐ Customer Retention ☐ API Calls per Business Transaction
Experience	☐ Top Customers by API usage ☐ Daily Support tickets per active users ☐ Customer Satisfaction Scores (CSAT) ☐ Net Promoter Score (NPS)	☐ Unique API Consumers ☐ Conversion Rate
Business Metrics for APIs		
	Day 1	**Day 2**
Revenue	☐ Monthly Recurring Revenue (MRR)	☐ Revenue vs Forecast ☐ Average Revenue per account (ARPA) Revenue by Acquisition Channel
Adoption	☐ SDK Adoption	☐ Feature Adoption
Churn	☐ Churn Rate	☐ Cohort Retention
Growth	☐ Growth Rate YoY ☐ Customer Acquisition Cost (CAC)	☐ Quota Alignment ☐ Net Sales Revenue ☐ Cost per Lead (CPL) ☐ Customer Lifetime Value (CLV, LTV)
Operations	☐ Support Metrics ☐ Beta vs General Availability	☐ Paid Support subscription rate By tier ☐ Cost of Infrastructure ☐ Cost of Outage ☐ Incidents per month

Index

A

B

D

P

W

Other Books You May Enjoy

If you enjoyed this book, you may be interested in these other books by Packt:

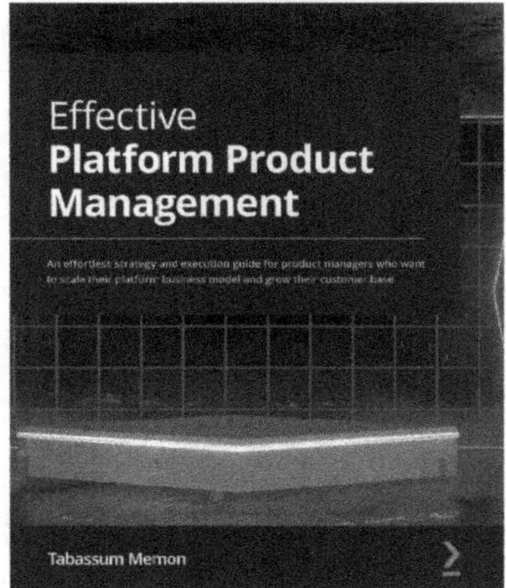

Effective Platform Product Management

Tabassum Memon

ISBN: 9781801811354

- Understand the difference between the product and platform business model
- Build an end-to-end platform strategy from scratch
- Translate the platform strategy to a roadmap with a well-defined implementation plan
- Define the MVP for faster releases and test viability in the early stages
- Create an operating model and design an execution plan
- Measure the success or failure of the platform and make iterations after feedback

Hands-On Low-Code Application Development with Salesforce

Enrico Murru

ISBN: 9781800209770

- Get to grips with the fundamentals of data modeling to enhance data quality
- Deliver dynamic configuration capabilities using custom settings and metadata types
- Secure your data by implementing the Salesforce security model
- Customize Salesforce applications with Lightning App Builder
- Create impressive pages for your community using Experience Builder
- Use Data Loader to import and export data without writing any code
- Embrace the Salesforce Ohana culture to share knowledge and learn from the global Salesforce community

Packt is searching for authors like you

If you're interested in becoming an author for Packt, please visit `authors.packtpub.com` and apply today. We have worked with thousands of developers and tech professionals, just like you, to help them share their insight with the global tech community. You can make a general application, apply for a specific hot topic that we are recruiting an author for, or submit your own idea.

Share Your Thoughts

Now you've finished *API Analytics for Product Managers*, we'd love to hear your thoughts! Scan the QR code below to go straight to the Amazon review page for this book and share your feedback or leave a review on the site that you purchased it from.

`https://packt.link/r/1-803-24765-7`

Your review is important to us and the tech community and will help us make sure we're delivering excellent quality content.

Download a free PDF copy of this book

Thanks for purchasing this book!

Do you like to read on the go but are unable to carry your print books everywhere? Is your eBook purchase not compatible with the device of your choice?

Don't worry, now with every Packt book you get a DRM-free PDF version of that book at no cost.

Read anywhere, any place, on any device. Search, copy, and paste code from your favorite technical books directly into your application.

The perks don't stop there, you can get exclusive access to discounts, newsletters, and great free content in your inbox daily

Follow these simple steps to get the benefits:

1. Scan the QR code or visit the link below

`https://packt.link/free-ebook/9781803247656`

2. Submit your proof of purchase
3. That's it! We'll send your free PDF and other benefits to your email directly

www.ingramcontent.com/pod-product-compliance
Lightning Source LLC
Chambersburg PA
CBHW082136210326
41599CB00031B/6005

* 9 7 8 1 8 0 3 2 4 7 6 5 6 *